# Food Safety for Beekeepers
## Advice on legal requirements and practical actions

### Andy Pedley MCIEH

This book is dedicated to Penny, soul mate of 44 years and patient reviewer of this text, sorting out many grammatical gaffs.

And to the many friends that I've made through beekeeping.

Northern Bee Books

**Food Safety for Beekeepers**
Advice on legal requirements and practical actions

© Andy Pedley

All rights reserved. No part of this publication may be reproduced, stored in a retrieval system, transmitted in any form or by any means electronic, mechanical, including photocopying, recording or otherwise without prior consent of the copyright holders.

ISBN 978-1-914934-37-7


Published by Northern Bee Books, 2022
Scout Bottom Farm
Mytholmroyd
Hebden Bridge
HX7 5JS (UK)


Design and artwork by DM Design and Print

# Table of Contents

Preface ................................................................................................................................5
The Law: ..............................................................................................................................8
   *Aim of the legislation* ....................................................................................................8
   *A food business?* ..........................................................................................................9
Business model ....................................................................................................................9
A step-by-step approach ....................................................................................................10
   *At the hive* ..................................................................................................................10
   *Harvest* ......................................................................................................................14
   *Extraction and processing* ...........................................................................................15
   *Labelling* ....................................................................................................................19
   *Discretionary labelling:* ...............................................................................................25
   *Storage prior to sale* ...................................................................................................26
   *A management system* ...............................................................................................27
Products ............................................................................................................................31
   *Hive products:* ............................................................................................................31
   *Other products* ..........................................................................................................36
   *Cosmetics* ...................................................................................................................38
   *Medicinal and Pharmaceutical products* .....................................................................39
Food Rooms ......................................................................................................................40
Food contact materials ......................................................................................................46
Food handlers ...................................................................................................................48
Cleaning ............................................................................................................................50
   *Cleaning equipment* ...................................................................................................50
   *Cleaning premises* ......................................................................................................52
Traceability .......................................................................................................................53
Allergens ...........................................................................................................................55
Second-hand equipment and recycling jars .......................................................................57
The Honey Regulations .....................................................................................................59
   *Definitions of different honeys* ...................................................................................60
   *Filtered honey and Baker's Honey* ..............................................................................62
   *Country of Origin* .......................................................................................................64
   *Floral Sources* ............................................................................................................64
   *Geographical origin* ...................................................................................................64
   *Enzymes* .....................................................................................................................66
   *Specific quality criteria* ...............................................................................................68
Registration ......................................................................................................................70
Food crime & Fraud ..........................................................................................................74

| | |
|---|---|
| Sampling programmes | 76 |
| *Analysis* | *77* |
| Enforcement | 78 |
| *Rights and powers of Officers* | *79* |
| *The inspection process* | *80* |
| GM crops | 82 |
| Due Diligence | 82 |
| Organic | 83 |
| Quantities and Weighing | 85 |
| *Weighing Systems* | *86* |
| *Weighing equipment* | *87* |
| Durability | 87 |
| Caveat | 89 |
| Appendix 1 | 89 |
| Appendix 2 | 94 |
| Appendix 3 | 97 |
| Credits | 98 |
| Endnotes | 99 |
| Alphabetical Index | 102 |

# Preface

Beekeepers are unusual, as their activity (which is often "just a hobby") is the production of food that may be sold to the public. Many will see the process through from the very beginning – making hives, sourcing bees, setting up apiaries, and managing the bees, right through all stages of harvest, processing, packing, labelling, to retail sale.

Training is usually through local beekeeping associations, who often use a curriculum set by the British Beekeepers Association; this focuses on the practical sides of beekeeping, husbandry and the like, and it is intended to lead on to further study with Modules, and even the National Bee Diploma, the pinnacle of the possible qualifications.

The BBKA's Module 2 assessment includes food production and hygiene: the "bee press" occasionally have articles or a mention.

There is little reliable guidance for beekeepers on the food production side, which is quite surprising. Although many beekeepers are hobbyists, with a few hives, and very limited sales, many have a number of hives. Of course, there are also bee farmers with many hives, producing significant quantities of honey and related products in commercial quantities. It's surprisingly easy to progress from a couple of hives to several, to having many!

There are often enquiries on social media about honey, processing, or legal requirements. The sometimes less-than-expert responses often perpetuate old practices or information, which were not always correct to start with. For instance, I've seen it stated that in order to label honey with its botanical origin, it needs to contain variously 40% through to 70% of that nectar; but neither of these numbers is mentioned in the UK legislation. Another misrepresentation is that you don't need to register as a food business if premises are used for less than five days in five consecutive weeks: this is based on the Food Premises (Registration) Regulations 1991. However, these were superseded by Retained Regulation (EC) 852/2004, which contained no such exemption; there's full information in the relevant paragraph.

Responsibility does, of course, lie with the food producer to inform themselves about the process and the governing legislation, and this book is intended to pull much of the information together into a single, practical guide. There are some assumptions here – the target audience is prime producers, beekeepers, and bee farmers – people who keep bees, process, pack and market honey. The text is intended to deal with all areas of law that apply to these activities. Beyond the scope of the book are importation and exportation, producing compound foods whether or not using honey as an ingredient, and the licensing legislation as it applies to intoxicating liquor. Although some references are made for completeness, the text cannot be seen as complete.

Food safety is sometimes seen as "food hygiene" and perceived in terms of cleanliness – all that is needed is for everything to be clean. However, there are many possible problems, many of them not obvious, and they will vary from food to food – a meat paste sandwich will present different hazards from a cheese sandwich. It is the food producer's responsibility to understand – and manage – any potential hazards associated with each product.

This book has been in the author's mind for some time – however, Brexit brought uncertainty. Prior to Brexit, UK legislation was harmonised with the EU, but the consequences of Brexit and subsequent trade deals with other nations might have brought about change, which would have rendered previous legislation irrelevant. However, although we left the EU on the 31st January 2020, for most of the UK almost all the EU legislation is "retained" and so remains in force – this is to avoid a "regulatory vacuum". The legislation will remain in place for the foreseeable future – although in September 2021[1] the UK Government announced a review of all legislation, it is likely to take some time. Additional reassurance is drawn from a survey[2] which suggested that the Food Safety legislation was important and the public did not wish to see it weakened in any significant way.

The intention is that this book will be accessible – a reasonable read, a useful and well-indexed reference source. It aims to give an understanding of the principles behind the requirements and practices, so that those principles can be applied to new situations not included here – and they are bound to arise. Illustrations are drawn from beyond the beekeeping world, where they are useful to demonstrate the underlying principles, or why particular steps are required.

"Ignorance of the law is no defence": it is incumbent on food businesses to know what the law is – but it would be easy to get bogged down in the minutiae of the legislation. It is the intention of the legislation that is important. A parallel can be made with road traffic legislation – the intention is safe roads and while most of us know enough to drive safely, few of us know the fine detail of the legislation. Where, for instance, in the legislation does it say that we drive on the left-hand side of the road? Which specific regulation imposes speed limits? But we understand the principles, that it is there to protect our safety, the environment, our ease of movement, and so on. I will endeavour to reference the specific legislation so that readers can review it for themselves and form their own interpretation of its actual meaning.

I should note that I have omitted some information that applies to very large-scale producers, in the interests of readability – few beekeepers are likely to use a tanker for transport, so including information would be redundant. But in the diversity of beekeeping, it may well be that some use bulk containers or tankers – if your business is exceptional, or out of the ordinary, then you may well want to review the actual legislation, research further, to obtain further advice or information.

The legislation is pragmatic, and it is there principally to protect the safety and quality of food – to ensure that, when we buy "honey" we get what we expect, the "natural sweet substance produced by *Apis mellifera* bees" rather than some factory-produced sugar syrup masquerading as honey. It is also there to ensure that "honey" is safe for us to consume – that it is free from harmful chemicals and bacteria, and that it has a reasonable shelf life.

Legislation requires enforcement – for the Road Traffic Acts, the Police, Traffic Wardens and the like. In the case of Food Safety, it is generally the Local Authorities who enforce registration and inspect premises and processes. In the case of County Councils, the Trading Standards function is undertaken by the County Council. Veterinary Medicines are overseen by the Veterinary Medicines Directorate, the Pharmaceutical legislation by the Medicines and Healthcare Products Regulatory Agency, Pesticide Regulations are controlled by the H&SE. And they all impinge on beekeeping. Here I'm focussing on Food Safety, so any mention of other areas of law is where they cross over to affect the safety of food, which they often do.

It is also important to note that marketing claims are controlled by the Advertising Standards Authority.

I should note too that this book is intended to apply to the law as it applies in England; there are separate, but very similar, laws that apply in the other nations of the UK, Scotland, Northern Ireland and Wales, and as our legislation is still based on the EU Regulations, there will be strong parallels in the EU. In the case of Northern Ireland, the NI Protocol means that to all intents and purposes, they continue to be in the EU.

I must also include a caveat – that the law is complex, and open to interpretation. Here, I have attempted to indicate what the legislation is, and how it applies to the beekeeping community. It has been done with care, considerable research, and the experience of 30 years a beekeeper, and 46 as an Environmental Health Officer, with a raft of food safety and other experience, including prosecutions, although my specialism was not Food Safety.

© Eric Flamant | Dreamstime.com

Often Enforcement Officers will refer to the actual legislation, and refer to relevant case law, when considering enforcement action. Encyclopaedias are available which are updated to reflect amendments to the legislation, relevant case law, and the like – local authorities will subscribe to these, and refer to them. Discussions take place on the fine nuances of legislation. Legal opinion may be sought … so, here I do my best to summarise the legislation accurately and comprehensively, but ultimately it is my view, my interpretation – others will disagree, and I may well be wrong. Only the Courts can interpret the law.

## The Law:

Law comes from a variety of sources; here we are mainly dealing with statutory law, law that has been passed by, or sanctioned by, Parliament. Despite Brexit, our food safety legislation is still based on European Law, "retained" to avoid a legislative vacuum, until new UK or England specific legislation can be brought in.

**Aim of the legislation**
The law aims to **"to ensure a high level of consumer protection with regard to food safety"** – and this is what we expect as consumers, that we are protected against harm from our food, that it is free from dangerous bacteria, viruses and other microorganisms, that it is free from harmful chemicals, including pesticides, heavy metal contamination and the like.

The law also ensures that what we buy is what we expect – in the case of honey, the sweet sticky substance that is produced by *Apis mellifera*, the European honey bee – and that it has not been adulterated with other syrups and sugars.

And the law imposes a duty on food producers to **"ensure that all stages of production, processing and distribution of food under their control satisfy the relevant hygiene requirements"**. A huge number of businesses and people are involved in the production of food – most of which will start life at a farm, then be transported, processed, packed, transported again to a wholesaler or warehouse and then on to the retail sale. Transportation may involve airfreight or shipping and international borders may be crossed. Each handler is responsible for the food while it is under their control.

The law applies universally to all food producers – the giant supermarket chains, the huge food factories, right through to the corner shop, the local baker, the stalls in the local market – and to beekeepers if they sell honey, or any other produce.

For beekeepers, the process may be entirely within our control – we assemble and site hives, manipulate the bees, harvest, extract, process and pack the honey, and usually sell it, so that duty lies with us.

## A food business?

> **Regulation 178/2002 on General Food Law, defines a food business as:**
> *"any undertaking, whether for profit or not and whether public or private, carrying out any of the activities related to any stage of production, processing and distribution of food."*

A lot of beekeepers would not consider themselves to be food businesses: "It's just a hobby", "I've only got a few hives" – but the moment that you start to sell food – and that is **any** food – you become a Food Business Operator (FBO), and you have responsibility to produce safe food, just as any other food producer. There may also be an obligation to register as a food business – I'll discuss this later in the book.

## Business model

A lot of beekeepers just produce honey, maybe creaming it and then selling it.

But it's quite common for the business to grow and develop – you get more hives, so find you need a way to deal with Baker's Honey, compound foods containing honey start to be produced and sold. Products such as Honey Mustard, Honey Marmalade, Honey Bread, Honey Cake, Honey Fudge are all compound foods involving ingredients and processing such as cooking.

When this happens, the legal requirements become more complicated: in particular, labelling, allergies, and durability become more important, and there will be a need to register with the Local Authority as a food business.

# A step-by-step approach

## At the hive

What we do at the hive, our husbandry of the bees, can impact on both the quantity and the quality of our crop. We need to be aware of this when working with the bees.

So considering what we do, and what we use, is important – what could affect our product, and what can we do to prevent (control) any harm?

### Smoke

Our smokers produce what we hope is a benign smoke – I usually use clean hessian sacking, but I have no knowledge of the toxicity of the smoke. The best I can do is to avoid materials that I know could be harmful, so shavings from tanalised (pressure-treated) timber, for instance, that would contain a variety of insecticides and fungicides are not used. Since I'm likely to inhale some of the smoke as I'm working, the effect of a lungful of preservative would be greater on me than on the honey, but it's worth being aware that smoke applied when working on the bees could potentially contaminate the product.

### Honeycomb

Honeycomb is – in effect – a food container, and must not contaminate the honey, so it needs to be stored carefully, where it does not become dirty or mouldy. Winter storage generally needs to be ventilated so that moisture and mould do not build up. Freezing is a good way to protect it against wax moth damage – at the time of writing there are no licensed treatments against wax moth.

© Katerina Kovaleva | Dreamstime.com

PDB, used as crystals that sublime, is a fumigant; both it and a similar chemical, Naphthalene, are lipophilic (soluble in fat), and so accumulated in beeswax when it was used as a fumigant of stored honeycomb. Although only weakly soluble in water, analysis has found PDB in honey.[3] Both chemicals have mammalian toxicity, could cause kidney damage, and are thought to be carcinogenic. There is no "Maximum Residue Limit" for either PDB or Naphthalene, so its presence in any food in any detectable quantity is not permitted. As noted, it has been found in honey. It is not licensed, and must not be used.

## Supers

When working at the hive, care does need to be taken to protect the supers, and so the honeycomb and honey – so the standard beekeeping practice of putting supers on a stand, or the inverted roof, to keep them off the ground is good practice.

Supers stacked on a roof to keep them uncontaminated when working at the hives.

## Feeding

Feeding the bees requires the use of wholesome products – typically granulated sugar has been used, and latterly fondant, Baker's Fondant, and premixed syrup solutions are commonly used. The feed that is given to the bees could affect the hive products, and care needs to be taken that it does not. Using proprietary products, specifically manufactured and marketed for use as a bee feed, should be fine, as should using food-quality granulated sugar.

Of course, care needs to be taken that the feed is given in such a way, and at a time, when there's no chance of it being stored in the supers and so contaminating the honey.

Some feeds have batch numbers, and noting these in the hive records would be best practice – in the unlikely event of a problem with a batch of feed, knowing the batch number used would enable you to check if you had used that batch and, if so, respond appropriately.

Using chemical repellents to clear the bees when taking supers off could also contaminate the produce, so using purpose-made proprietary repellents is important.

In 2008 the National Residues Monitoring Programme of the Irish Food Standards Agency's Pesticides Control Services discovered that pork was contaminated with dioxins and PCBs. Subsequent investigation showed that the contamination was from the animal's feed, which had been contaminated by the hot gases used in its drying.

The incident resulted in 30,000 tonnes of product being destroyed, as well as 170,000 pigs and 5,700 cattle, with a cost to the Irish Exchequer in excess of 120 million.

### Treating the bees

Bees are now regularly treated with medications, "Tonics", "Hive Cleaners" and the like. These have the potential to affect the honey.

It is important to only use licensed products, to use them strictly in accordance with the manufacturer's instructions, and to keep the required notes for five years.

Feeding bees with sugar syrup, fondant or pollen substitutes should not have any adverse effect on the honey, other than the possibility of the honey becoming contaminated by feed, and good beekeeping practice is not to feed when the supers are in place.

Protection against wax moth is probably a pesticide treatment rather than a veterinary medicine, so different legislation applies and is administered by the Health and Safety Executive. Anything that is applied – in any way – to prevent or reduce wax moth damage almost certainly falls within the definition of a pesticide.

Sulphur Dioxide gas, from burning sulphur-impregnated strips, has also been used as a fumigant, although this also does not have a licence, and so cannot be used.

Another fumigant is Acetic Acid, which is said to be antibacterial as well as effective against the wax moth. Again, it is not licensed so should not be used.

To summarise – unless there is a licence for the product that you are using for the specific use, it should not be used. The product should be used strictly in accordance with the manufacturer's instructions, and in the case of Veterinary Medicines, appropriate records should be kept.

Note that licences are **not** transferable – that some proprietary Oxalic Acid-based treatments are licensed as Varroa treatments does **not** mean that Oxalic Acid BP can be used.

### *Hive protection*
The exterior of wooden hives, particularly, are often treated with paint, oil, or wood preservatives. Although the hive boxes themselves are not food contact materials, there is a small risk of contamination, and this needs to be borne in mind when selecting the treatment. Paint, and oils such as linseed oil, will generally be innocuous, but beware preservatives, especially those containing insecticides and fungicides.

# Harvest

Harvesting is when we come into direct contact with the food, and so is the time when we need to take most care. There are several stages.

*Transportation*
So that glorious moment comes when the harvest is to be taken off, and the supers – after clearing, of course – need to be taken for extraction.

Many beekeepers will use one or more vehicles – a barrow, car boot, or van, and steps need to be taken to protect the honey comb from contamination – car boots, for instance, may have soil from boots, hairs from pet dogs, debris from trips to "the dump", etc., and are often carpeted, so difficult to clean out.

So, making sure that the vehicle is lined with something clean – plastic sheet or clean, disposable paper, for instance – is good practice. You'll probably want to do this to protect the vehicle from the odd drip of honey / smudge of propolis anyway, and it should be done whenever supers are transported.

A small point, but lining the vehicle would be particularly important in top bee space hives, as the bottom of the frames will align with the bottom of the super and so be more vulnerable to picking up debris from the vehicle floor.

There's guidance on the FSA website[4] and there are specific legal requirements which include:
- that the vehicle and any containers that food is transported in are clean and used just for transporting foodstuffs;
- that there's effective separation of food and non-food products as necessary.

Be aware that I've omitted several provisions as they are unlikely to apply to the majority of beekeepers – get further guidance if you use bulk containers or a tanker, for instance.

*Storage prior to extracting*

It's likely that the supers will need to be stored, before the honey is processed: this may only be for a short time. During this, the honey is very vulnerable to harm: uncapped honey may absorb moisture, and pests – such as ants, mice, wasps, and even robbing bees – may discover the stash and give it their attention. I like to minimise the storage time (ideally remove from the hives one day and extract the next) to minimise this period of vulnerability. The supers are best stored in the food room, protected from the ground by a tray or sheet of some kind, and covered to protect them against dust, moisture, and wasps. The room needs to be bee- and wasp-proof.

## Extraction and processing

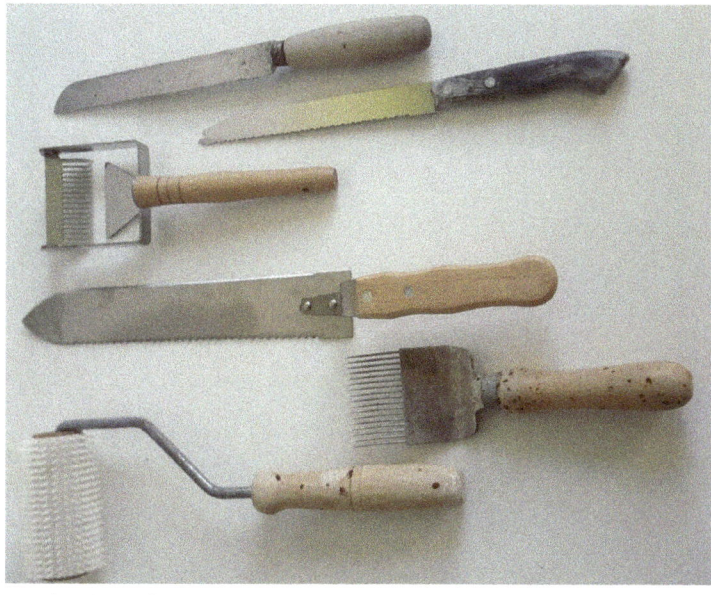

A selection of uncapping tools

This needs to be done indoors, so you'll be working in a food room of some kind. This may be your domestic kitchen, or some kind of special facility – some beekeepers have "honey houses", and some associations have Extracting Rooms that are available for their members to use. Some will use a friend's extracting room, or access to a commercial kitchen is available, and the extracting, etc. is done there.

There are requirements for "food rooms", but if you are using domestic premises, then there are different standards that apply, and there is a section on Food Rooms dealing with the requirements for both.

Everything used during extracting needs to be clean and suitable for use (see the section below), and this is the moment when the honey begins to be exposed to a new potential source of contamination, the food handler – again, see the section below.

*Storage prior to bottling*

Honey is likely to be stored in honey buckets, which are available from beekeeping suppliers, or direct from the manufacturers. I've known associations purchase these in bulk and distribute them to their members.

They'll be made of plastic, which becomes a food contact material, so they need to comply with the requirements, which will be evidenced by the cup and fork symbol moulded into the bucket and the lid.

**Heating to liquefy**

Stored honey will crystallise, sooner or later, and may need to be liquefied to enable it to be strained and packed into jars.

There are proprietary "warming cabinets" available from beekeeping suppliers, and beekeepers are adept at creating their own improvised systems – there are instructions online.

| Temperature | Time to generate 40 mg/kg HMF |
|---|---|
| 30°C / 86°F | 250 days |
| 40°C / 104°F | 50 days |
| 50°C / 122°F | 10 days |
| 60°C / 140°F | 2 days |
| 70°C / 158°F | 0.4 days (10 hours) |

A table showing the development of HMF at different temperatures. Courtesy of Quality Services International America.[5]

A thermostatic control. This one was designed for use with an aquarium. Wiring it safely requires electrical expertise.

Warming honey does raise the level of HMF – however, if the temperature is not allowed to go too high, then the research indicates that it would take many days to exceed the legal limit of 40 mg/kg.

Different honeys respond to heating differently and there is no hard and fast rule on how rapidly HMF develops, so err on the side of caution and allow a wide safety margin.

Using a thermostatic control to limit the temperature, and a timer to turn the heater off after a set time – both "critical controls" in the HACCP referenced below – I find that with my set-up, 24 hours with the thermostat set at 42.5°C is sufficient to re-liquefy a 30lb honey bucket. Others suggest that 32°C, 38°C, or 40°C is sufficient, and the lower the better.

I can be confident that the HMF has not built up to an illegal level thanks to a variety of research, showing the rate of HMF development over a range of temperatures. It is very slow at lower temperatures. As "due diligence", I take care not to overheat the honey (hence the thermostat) or heat it for too long (hence the timer), and my records include the maximum temperature and the duration of the heating in the batch record for the honey.

Apparently microwave energy can adversely affect enzyme activity, so microwave ovens should not be used to liquefy crystallised honey.

Less information is available about Diastase: however, research[6] states that Diastase decreased to below limit values when honey was heated to 75°C for 180 Minutes, and the decay is much slower at lower temperatures.

*Straining*

An essential for beekeepers is straining the honey, to remove, mainly, odd bits of wax, bees' anatomy, etc.

A double strainer in use - the upper strainer is a course mesh, and the lower much finer. © Airborne77 | Dreamstime.com

Strainers – sieves – come in a variety of shapes, sizes, and made from a range of materials. Critically the mesh size will vary too, and many beekeepers will use several different strainers to progressively remove smaller and smaller debris.

The Honey Regulations prohibit straining honey to such an extent that the pollen is removed. The mesh size is not defined, but may be dependent to some extent on the crop being processed. Pollen grains vary hugely in size, but are typically 100 microns or less (about the diameter of a human hair) so any strainer that has a coarser mesh than this is acceptable – commercially I understand that straining media of 80 microns are in use, and the majority of pollen would pass through this. Natural fabrics (muslin) would generally exceed this.

Practically, for most beekeepers, honey is strained at or close to room temperature, and its viscosity means that it will not pass through very fine strainers in a reasonable period of time. The commercial example given above was a large production facility, where the honey was processed at 55°C, and pumped through the filter media.

The filter media, whether a sieve, strainer, or filter cloth, are food contact materials and come into intimate contact with the honey. They must be food-grade, and handled with care to prevent them becoming contaminated.

References on the internet, and in "older" beekeeping publications, suggest that the use of non-food grade materials is acceptable (I've seen net curtains and ladies' tights referenced). These would not be food-safe, and must not be used.

Filter media need to be washed with food-safe detergent, and cloth / fabric needs to be hand-washed, and not put in the washing machine with other clothing.

*Ripening*

This is the last stage before packing and in my process immediately follows straining. In fact, it is a bit of a misnomer – the honey is not actually ripened, but allowed to stand for 24 hours in the bottling tank so air can rise to the top, which improves the appearance of the honey in the jar. If this is not done, the air will rise and form a froth on the top of the jar. Though completely natural and harmless, it mars the appearance and so spoils the customer experience.

The bottling tank and honey gate need to be food grade, as well as the honey buckets. As with extractors, they used to be made out of tinplate with lead-soldered seams, but now are of plastic (look for the cup and fork symbol moulded in) or stainless steel.

*Packing*

Typically, jars are made of glass, but plastic jars are also available.

The jar, and the lid, its seal, and any lubricant on the seal will all be food contact materials and must be food-safe. Jars and lids sold for use with food will comply with this, but containers not produced specifically for use with food should be avoided, or the manufacturers consulted to determine suitability.

Other presentations of honey (comb, or cut comb) may have different packaging – sections will be in their wooden or plastic housing, and it's wrapped in a plastic or cellophane container. The wrapper is a food contact material, so needs to be food-safe.

Honey may be sold with, or in, a decorative or ornamental jar – craftsman potters may use special glazes, and as honey is acidic, it can leach chemicals out of the glaze. Care needs to be taken that the potter is aware of this and that their selection of materials in the glaze does not include heavy metals. The concern here is, lead and cadmium (gives a red colour) – and there are specific regulations – Materials and Articles in Contact with Food (England) Regulations 2012 – that apply. Research shows that the ceramic industry is well aware of the potential hazards, and the majority of glazes are free of both chemicals. However, the Regulations require specified tests to be carried out unless it can be shown that that the materials used did not contain either chemical – this is likely to involve a declaration by the producer that they had not used such glazes.

There can be a complex interaction between food and the materials that it is in contact with, and hence the need for strict regulations. Often the interactions would be less than obvious, especially in the case of plastics, which may contain a variety of additives: in addition to potentially being toxic, they could also be carcinogenic or "endocrine disruptors" – chemicals which interfere with the endocrine (hormonal) systems that regulate our bodies. Such a chemical is Bisphenol A (BPA). It may be present in polycarbonates, epoxies, and flexible PVC – its use can be either as a plasticiser or as an antioxidant. It mimics oestrogen and in 2017 it was listed as a substance of very high concern due to its properties as an endocrine disruptor.

 For this reason, regulations require that packaging, made from, or lined with plastics, and intended for food contact, are marked either with the words "for food contact" or with a "cup and fork" symbol or similar.

The Food Contact Materials regulations impose specific requirements on packaging made from Regenerated Cellulose packing, plastics and recycled plastics, Epoxy derivatives, and Vinyl Chloride.

## Labelling

There are, of course, requirements that food is labelled and the content of the label is specified, some information is mandatory, and some optional at the discretion of the producer.

### *Mandatory information*
Mandatory information must be in the "same field of view" and on a plain background, and with good contrast.

For honey, the following information is mandatory:

- the product name
- the name or trade name and address of the producer or responsible food business operator
- the country or countries of origin
- any special storage conditions

- a best-before date
- a lot mark
- the weight

Dealing with each in turn:

### *The name (description of the food)*

The word "Honey" is a "Reserved Description" and can only be applied to the substance that complies with the definition in the Honey Regulations.

The Regulations also allow the use of some supplementary descriptions that explain the source from which the honey is obtained (for example, blossom, honeydew), the processes by which it's extracted (for example, drained, extracted), its presentation (comb, chunk honey) and where it came from – those are all clearly permitted by the Honey Regulations.

It's worth making the point that there is no legislative provision for the use of other descriptions – "Pure", "Raw", "unfiltered", "Natural", "Artisan", "un-pasteurised" and the like: there is more on this below.

A display showing a wide variety of label designs

### *Name and address of the producer*

The name and postal address of the producer must be on the label.

The address can be abbreviated down to just a house number or name and the postcode in the interests of brevity and saving space on a small label. So the publisher's name and address – Northern Bee Books, Scout Bottom Ln, Scout Rd, Mytholmroyd, Hebden Bridge, HX7 5JS – could be abbreviated just to: Northern Bee Books HX7 5JS.

A limited company is likely to give its Registered Office or Principal Place of Business as its address.

From the 1st October 2022, the name and address of the Food Business Operator must be in the UK; if they are not in the UK, then the importer's name and address needs to be given.

*Country of origin*

The label must show the country of origin of the honey. It is not sufficient for the description to be "Oxfordshire Honey" – there needs to be a specific statement, "produce of the UK" or something similar.

If the honey is blended and from more than one country, then it needs to be labelled appropriately:

- "blend of honeys from more than one country" (or similar wording)
- "blend of EU honeys"
- "blend of non-EU honeys"
- "blend of EU and non-EU honeys"

Bear in mind that the UK is no longer a member of the EU.

After 1st October 2022, you must use "blend of honeys from more than one country" (or similar wording) if you decide not to list each country of origin.

According to the FSA Website, in Scotland, slightly different requirements will apply, and the options for labelling origin include using:

- "blend of honeys from more than one country" (or similar wording)
- reference to the trading bloc of the countries of origin (for example, "blend of EU honeys", "blend of non-EU honeys" or "blend of EU and non-EU honeys")

If you continue to use trading bloc or EU terms, you must ensure your label is accurate. For example, a blend of UK and French honey placed on the market in Scotland would need to either:

- list both countries ("blend of honeys from Scotland and France") or
- use the words "blend of EU and non-EU honeys".

This is because the UK is no longer part of the EU.

And in Northern Ireland, if you place a blend of honeys from different countries on the NI or EU markets, you must accurately reflect that GB honey is no longer EU honey and use one of the following terms:

- "blend of EU honeys"
- "blend of non-EU honeys"
- "blend of EU and non-EU honeys"

### Special storage conditions

As honey is an ambient food, it does not need to be kept refrigerated. However, it is important that the jar is hermetically sealed to prevent moisture being absorbed, and the honey from fermenting, so the label could include a statement, "Store with the lid screwed on tightly" or similar.

Other products may need more specific storage instructions.

### Durability

The durability – best before date – need not be printed on the label: an indelible marking on the jar, clearly handwritten, could be sufficient.

### A batch or lot number

This is important for traceability; it can be printed on the label or could be written, clearly, in indelible ink on the jar itself. It must be clear and legible.

### Weight

Honey is almost always sold "pre-packaged" – put into its container before sale. Other rules apply for non-pre-packaged food sales.

Honey is sold by weight, not volume,[7] and the weight in metric units must be on the packaging. The figure to give is the Nett Weight (just the weight of the product) rather than the Gross Weight (the product and the packaging).

Packs of less than 5 grams nett weight need not be marked with their weight.

The Metric system must be used, and Imperial units (pounds and ounces) can be used after the metric weights, and in the same size font or smaller if desired. Post-Brexit, this is under review.[8]

Honey in containers over 10g nett weight does not need to be marked with the weight, but only if they are not sold by retail, and if they are sold with accompanying documentation giving the weight.[9]

The size of the lettering for the weight mark is regulated – the following lettering sizes are to be used:

| Nominal quantity and unit of measurement | Minimum height of words or figures | Font size (Arial) |
|---|---|---|
| exceeding 1 kg | 6 mm | 50pt |
| exceeding 200g but not exceeding 1 kg | 4 mm | 24pt |
| exceeding 50g but not exceeding 200 g | 3 mm | 20pt |
| not exceeding 50g | 2 mm | 12pt |
| exceeding 1 L | 6 mm | 50pt |
| exceeding 20cl but not exceeding 1 L | 4 mm | 24pt |
| exceeding 5cl but not exceeding 20cl | 3 mm | 20pt |
| not exceeding 5cl | 2 mm | 12pt |

It is of course an offence to package less than the declared weight, but providing a surplus could be seen as an offence too – though there is a defence *"for the person charged to prove that the excess was attributable to the taking of measures reasonably necessary in order to avoid the commission of an offence in respect of a deficiency in those or other goods"*.[11]

## Other foods labelling requirements

In addition to the requirements above, all other foods need to have the following information on the label:

### Allergen labelling

See the chapter about allergens and labelling – accurate labelling is essential for compound foods.

### Organic

It should also be noted that the word "organic" has a very specific and legally defined use, and **MUST NOT BE USED** unless you meet the criteria. Please see the separate chapter.

### Nutritional value

In general, it is mandatory to provide information about the nutritional value of a food on the label – at a minimum:

- energy value
- amounts of fat, saturates, carbohydrate, sugars, protein and salt

However, there are exemptions in EU Regulation (EU) No 1169/2011 ANNEX V, including:

- unprocessed products that comprise a single ingredient or category of ingredients; and
- food, including handcrafted food, directly supplied by the manufacturer of small quantities of products to the final consumer or to local retail establishments directly supplying the final consumer.

The first exemption certainly applies to honey, and the second applies to honey and could apply to some products, so long as it's supplied by the manufacturer direct to the customer or to a local retailer. But, as with registration and primary production, there is no definition of small quantities or local.

### *Ingredients*

For honey, there is no need to list the ingredients as it is a single ingredient food, and the name clearly identifies it.

However, for all pre-packaged foods with ingredients, there must be a list of the ingredients, headed "Ingredients" and they need to be listed in order of weight.

Sometimes it will be necessary to give a quantitative ingredients declaration (QUID) – there's guidance on the FSA website. So if the name of the food includes the ingredient – for instance, a product marketed as "Honey Mustard" – it would require the quantity (%) in the product to be given on the label.

There are other circumstances where a QUID is required – for instance if the ingredients are not typical for the product (for instance if pork has been used in a lasagne, which is usually made with beef), or if there's an illustration that suggests the presence of the ingredient in the food (so a picture of blackberries, for instance, could require the percentage of blackberries in the product to be shown).

There are specific requirements if the ingredients include sweeteners or sugars, aspartame and colourings, liquorice, caffeine, or polyols, and of course any allergens present must be listed and emphasised as above.

### *General labelling provisions*

Apart from the minimum height for the weight marking, lettering for the other mandatory information on labels must have an "x" height of at least 1.2mm, or 0.9mm if the largest surface area of the packaging is less than 80 square centimetres.

Note that there is not consistency between different fonts – but for Arial, a font of 10 points gives 1.2mm, 12-point gives 2mm, 20-point gives 3mm, 24-point gives 4mm, and 50 gives 6mm. Labels must be difficult to remove or damage. Self-adhesive labels are available with different types of adhesive, permanent and temporary. The permanent type is preferable – temporary peel off very easily. Ink should be water-fast (inkjet printers are unlikely to be good enough). Tie-on tags and similar are likely not to be good enough on their own, though if the mandatory information is on an adhesive label, additional information could be given on a tie-on tag or such.

If you sell in bulk, for instance to someone else who is going to package the honey and sell it, or is going to use it as an ingredient, you don't need to label the packaging, so long as the documentation you supply with the honey includes all that you would put on a retail label – your customer would label the products with their name and address, and their batch records would record that you were the source of the honey in that particular batch of jars / honey cake, etc. For traceability, they need to know the name and address of the supplier.

**Discretionary labelling:**
Additional information, email addresses, websites, telephone numbers may be added, but they **cannot be substituted** for a postal address, which must always be present.

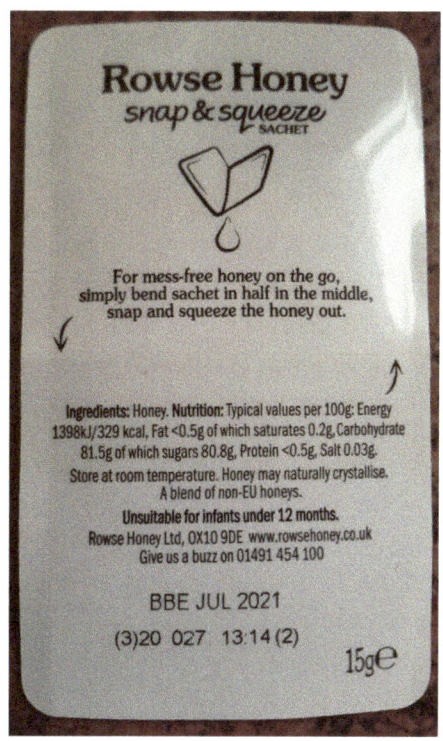

A fully compliant package of commercially produced honey - it includes an Ingredients list (though this is not necessary); nutritional values (again, not necessary); storage instructions (also optional); a comment about crystallisation (not essential); legally required information about the source of the Honey (a blend of ...); the warning not to give honey to infants (not obligatory, but due diligence and in compliance with the Honey Packers Association's Code of Practice), the name and address of the packer (the address abbreviated down to just the name and postcode - the postcode is likely to be specific to the Company's address, if not it would need to be with the Unit Number etc); supplementary information - website and phone number (both optional); a BBE date, a batch number, and finally, the weight. The optional letter $e$ indicates that the packaging is packed using the Average, rather than the Minimum system.

### *Warning not to give honey to infants under 12 months of age*

This is not mandatory, but the British Honey Importers and Packers Association (BHIPA) Code Of Practice specifies that the label should contain the following wording: "Unsuitable for children under 12 months" – and it could be argued that this is required under the due diligence requirements.

This is because honey can contain the spores of Clostridium Botulinum,[12] a bacterium that causes the food poisoning, Botulism. The spores cannot develop in honey, due to the high Water Activity, presence of hydrogen peroxide, and its high acidity. However, when honey is consumed by a child with an immature gut microbiome, the spores can develop and the resulting bacteria produce the Botulism Toxin, which leads to infantile botulism, or "floppy baby syndrome" – a very serious condition, fatal in 5% to 10% of cases. As a child ages, their gut flora becomes more complex and the gut becomes unsuitable for Clostridium Botulinum to propagate, so honey is safe for older children and adults.

### *Security seals*

Specially-made security seals are available to be put on containers – these are not obligatory, but are an excellent idea and certainly good practice.

I have known customers open jars to smell, and taste, honey that's displayed for sale, and so remove the lid and prejudice its purity, etc. The presence of a security seal protects you both against fraud, but also sends a message to potential customers that the jar is sealed and not to be opened.

Seals in a variety of designs are available from beekeeping suppliers, and are simple to apply.

Security Seals came into common use following the case of Rodney Whitchelo,[13] an ex- Scotland Yard detective, who attempted to blackmail the food manufacturer Heinz by contaminating some jars of baby food with glass shards and razor blades. A security seal would have made the tampering obvious.

### *Storage prior to sale*
Once packed (bottled), the honey is well protected against harm, but it still needs to be protected. Ideally storage will be a cool location, to prevent HMF building up, and it must be pest-free. Glass jars can be broken, so care is needed when moving or handling the jars. Dirt on the outside of the jar will mar its appearance and so its marketability, as will damage to the label.

## A management system

The legislation requires that there is a "Management System" in place to ensure that food is safe and wholesome, and that potential hazards are managed.

That sounds very grand, and perhaps even intimidating; but as beekeepers extract and pack their honey and other produce, they are already using a system which they have probably learned from other beekeepers, or in this day and age, YouTube and other sources online. Many will include many of the steps and precautions that are taken above, and they may not realise it but the process they use is, in effect, a system which probably includes all or most of the steps required to keep honey clean, safe, and free from contamination.

The best practice method of managing the hazards, and the industry standard one, is a process known as "Hazard Analysis and Critical Control Point" or HACCP for short, and the steps that I've outlined above are based on my HACCP, which is reproduced in Appendix 1.

Guidance on this on the Food Safety Authority of Ireland website indicates that low-risk food businesses do not need to develop an HACCP system, and allows businesses to follow good practice where hazards and controls have been identified.

The key steps in an HACCP process are:

(a)  identifying any hazards that must be prevented, eliminated or reduced to acceptable levels;
(b)  identifying the critical control points at the step or steps at which control is essential to prevent or eliminate a hazard or to reduce it to acceptable levels;
(c)  establishing critical limits at critical control points which separate acceptability from unacceptability for the prevention, elimination or reduction of identified hazards;
(d)  establishing and implementing effective monitoring procedures at critical control points;
(e)  establishing corrective actions when monitoring indicates that a critical control point is not under control;
(f)  establishing procedures, which shall be carried out regularly, to verify that the measures outlined in subparagraphs (a) to (e) are working effectively; and
(g)  establishing documents and records commensurate with the nature and size of the food business to demonstrate the effective application of the measures outlined in subparagraphs (a) to (f).

> It's worth making the point, regarding documents and records, that EU Regulation 852 / 2004 states:
>
> *Food business operators are to keep and retain records relating to measures put in place to control hazards in an appropriate manner and for an appropriate period, commensurate with the nature and size of the food business.*
>
> This is open to wide interpretation, but the "nature" of honey production is low risk, and many beekeepers are small-scale, so it could be argued that there is less need for detailed records – documents – than a much larger business, producing a high-risk food. With some exceptions, it is for the food business operator to decide what records they need to keep.

This is usually done, in practice, by plotting the product's journey – in the case of honey, from the hive to the consumer. For the beekeeper, this generally includes everything from the hive to the jar.

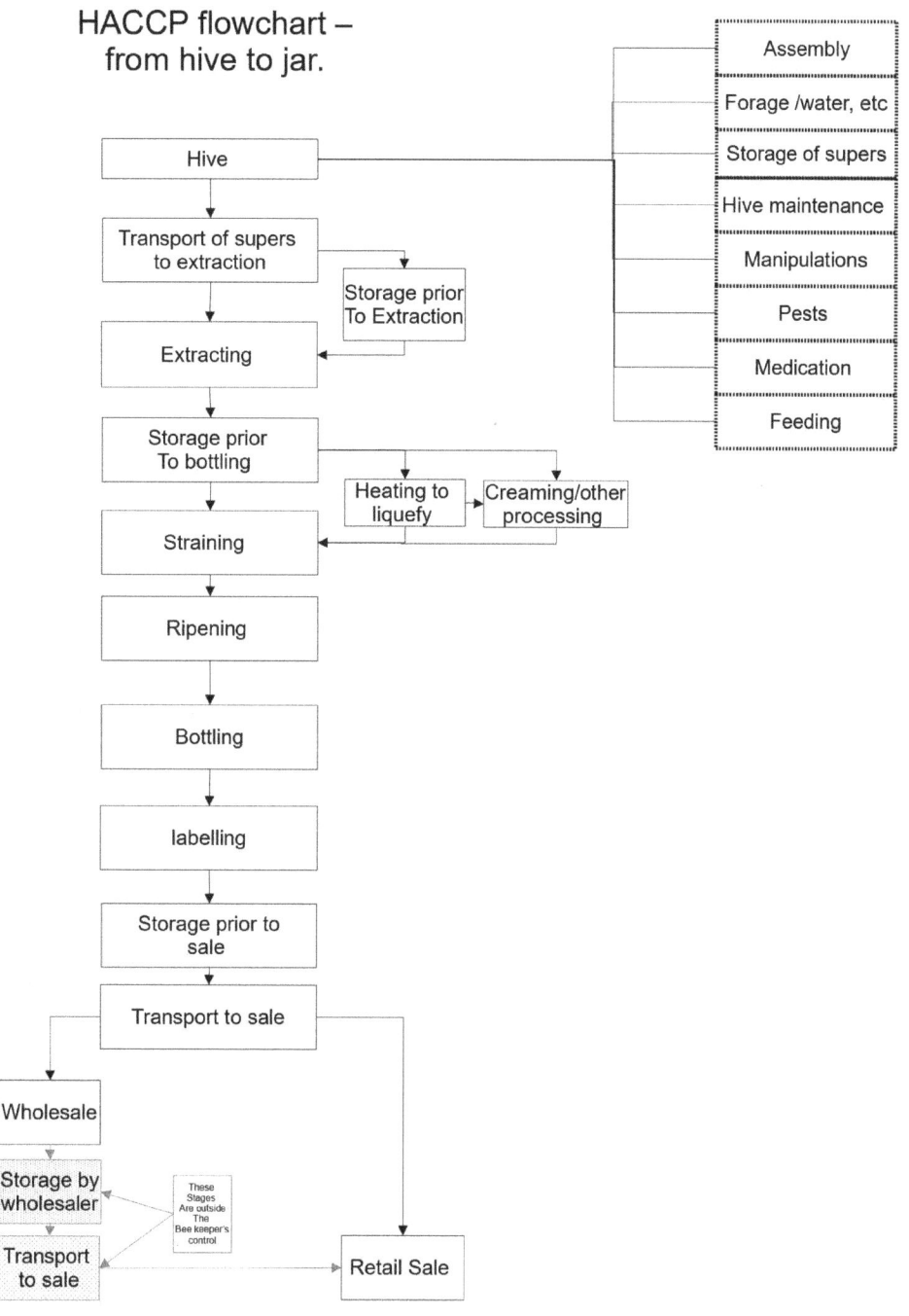

The next stage is to review this flowchart, to identify what could go wrong and where. This is the "hazard analysis".

In the case of honey, and its processing, the following could be present and be potentially harmful:

- water,
- medications and treatments, including antibiotics,
- hive cleansers,
- debris – bits of bee, wax, dirt, wax moth faeces,
- pests,
- heat,
- age,
- heavy metals,
- contaminants from air or water pollution, and
- agrochemicals from forage or elsewhere.

This list may not be exclusive.

Decide what you already do to prevent these potential harms, and when they are in process. These are critical controls and control points. Consider, are they adequate? How would you know if there is a failure, and what more could or should you do to prevent them? Add these to the process.

> An example here would be excessive water content in your honey, which will lead to spoilage – fermentation – and the Honey Regulations impose a limit on the water content.
>
> **What controls exist to prevent an excessive water content in the honey?**
> The first is that you only extract from fully-capped frames, or use the "shake test" before extracting (this is not especially reliable!); the second is that you keep the room as dry as possible when processing honey – if the floor has been mopped, it's been allowed to dry so the RH of the room does not affect the honey. Third, you keep the honey hermetically sealed – in honey buckets with the lid clipped down, or in sealed jars.
>
> **What is an excessive water content?**
> This would be the 20% figure from the Honey Regulations.
>
> **How would you know if there's an excessive water content?**
> You can use a refractometer to measure the water content.
>
> **Finally, should the honey fail the test, what can be done?**
> Since the Apimondia definition of Honey Fraud includes artificially desiccating the honey, that is not an option; however, the honey could either be sold as "Baker's Honey", fed back to the bees, or used as an ingredient in a honey product – honey cake, or reserving it for personal consumption, are all possible ways to use the honey.

There is a template in Appendix 1, but please note that this is an example only and each beekeeper needs to produce their own, reflecting their own circumstances, processes, equipment and produce. Finally, decide what you will do if a batch of honey is substandard, and document that.

## Products

**Hive products:**
It's generally reckoned that there are six possible products from a hive – these are:

| | | |
|---|---|---|
| Honey | Pollen | Royal Jelly |
| Beeswax | Propolis | Venom |

Beekeepers are innovative, however, and may well develop new products. I know of one beekeeper who sold Drones to a restaurant, and I've seen "Bee Bread" on sale online recently too.

Some of these may constitute "Novel Foods" – there's a definition of "novel food" on the Food Standards Agency (FSA) website as follows:

> *any food that was not used for human consumption to a significant degree within the (European) Union before 15 May 1997, irrespective of the dates of accession of Member States to the Union.*[14]

### *Honey ...*
... is defined in the Honey Regulations (England) 2015, as:

> *... the natural sweet substance produced by Apis mellifera bees from the nectar of plants or from secretions of living parts of plants or excretions of plant-sucking insects on the living parts of plants which the bees collect, transform by combining with specific substances of their own, deposit, dehydrate, store and leave in honeycombs to ripen and mature.*

... although there are other slightly different definitions elsewhere.

Honey of all colours on display at the National Honey Show

In addition to Sugars and Water, it contains a variety of enzymes, HMF, types of acidity and pH, insoluble solids, organic acids, proteins, vitamins, minerals, volatile and semi-volatile compounds and polyphenols. It has antioxidant and antimicrobial properties.[15] It will contain small amounts of pollen.

The anti-microbial properties that give it its legendary shelf-life are a result of:

- the high osmotic pressure – in food technology terms, the "Water Activity" or $a_w$
- the presence of hydrogen peroxide in the honey; this is itself a result of the enzyme Glucose Oxidase being present
- the acidity – typically the pH is somewhere between 3.4 and 5.1 depending on a number of factors.

The types of sugars are dependent on the forage crops that the bees have been to – and these change the properties of the honey. One of the most obvious is the crystallisation of the honey. Canola (Oil Seed Rape) honey is famous (notorious) for crystallising very rapidly, a consequence of its low fructose to glucose ratio: it needs to be processed quickly at the end of the flow, and requires harvesting and extracting during the flow to avoid problems with it granulating in the honeycomb.

© Ann Chillcott

Some honeys are thixotropic, notably that from heather – this requires liquefying before it can be extracted, using a centrifugal extractor, or it may be pressed out of the honey comb. Liquefying involves agitating with a "honey loosener" on a large scale, or a "perforextractor" on a smaller scale, and the honey remains fluid for about 15 minutes.[16]

In the UK regulations, honey is specifically the produce of the European Honey Bee, Apis mellifera – the international Codex Alimentarius[17] is not so specific and other bee species' produce also meets the definition.

Honey is an "ambient" food – it does not require refrigeration or freezing to prevent spoilage, so I do not deal with issues such as maintaining and monitoring a cold chain. It is conceivable some compound foods made by beekeepers may well require refrigeration – for instance, the shelf life of a honey cake could well be extended by refrigeration.

In the UK, there are no hazards presented by forage – research by the FSA confirmed that even bees foraging on ragwort and borage, both of which contain a toxin, do not produce honey with any significant toxin.[18] [19]

There are examples of honey across the world that are toxic – some rhododendron honey contains Diterpenoids, which are hallucinogenic, causing "mad honey disease" and in New Zealand, where honeydew, from the native Tutu bush can contain tutin – a remarkably potent toxin. Other toxins can affect honeys.[21]

### Beeswax

Beeswax, refined and moulded, ready for sale.

Wax is a natural secretion of the bees, and of course they use it to build honeycomb – the immensely versatile nest of the honeybee colony. They use it for raising brood, storing pollen and honey, and ripening nectar.

It is comprised mainly of esters of fatty acids and various long-chain alcohols.

Beeswax is edible, having similarly negligible toxicity to plant waxes, and is approved for food use in most countries and in the European Union under the E number E901 – it's a glazing agent. It's also known as *Cera alba*.

It may also be used as "bone wax" in orthopaedic surgery, and in cosmetics such as hand creams and lip balms.

---

*Beeswax for food use is more valuable than lower grade products – and deserves special treatment. In their September 2019 journal, Bees for Development tell of beeswax that has been contaminated with Permethrin, possibly as it had been filtered through Pyrethroid-treated mosquito nests. This has severely devalued the crop, and demonstrates the importance of only using purpose-made equipment, traceable to reputable suppliers, when processing any food product.*

*It also demonstrates the importance of sampling, to ensure that the product is free from contamination and suitable for its intended use.*

---

Much beeswax is not used for food, but where it is produced for food use, or bought for food use, then it needs to have been processed and stored as a food.

Recently beeswax is used, often in conjunction with other oils, in the manufacture of "food wraps", so although it is not being consumed, it is a food contact material. If it is to be used for such use, it needs to be of the highest cleanliness and purity. See below.

I've also seen "Beeswax Melts" on sale online – these appear to be a blend of beeswax and other oils or waxes, with added scents, and they are intended to be put into a vaporiser to liberate the odour in the same way as an incense burner does. These are dealt with below.

*Pollen*

Bee with full pollen baskets, collecting pollen from Willow. © Dave Massey | Dreamstime.com

Bees collect pollen; it is high in protein and provides important amino acids in the diet; it's used principally in the brood food, to nourish the larvae prior to their being capped.

Bees carry their pollen loads in "pollen baskets" on their back legs, and it will also be on the hairs on their bodies; it is naturally present in honey.

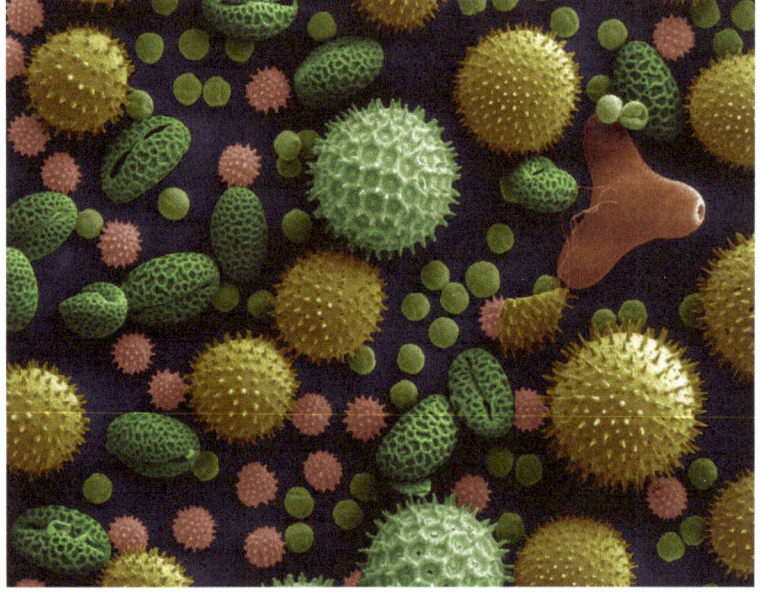

Above: A colourised scanning electron microscope image of pollen grains. Public domain, by consent of Dartmouth Electron Microscope Facility, Dartmouth College

Right: Pollen packed and ready for sale.

In addition to being the bee's principal source of protein, pollen is high in lipids (fats), and different pollens will contain different amino acids, so collecting pollen from a range of different plant species will help the bees acquire all the amino acids that they need for their nutrition. Pollens are very distinctive – different sizes, shapes, patterns, and these are unique to the original source. Melissopalynology – the study of pollen in honey – can identify the botanical origins of the honey, and so its geographical source.

Pollen is an unusual crop in the UK, and the method of collection leaves it vulnerable to contamination – a "pollen trap" is fixed to the front of the hive, and the bees have to pass through it as they return. There's a grid on the front of the trap, and the pollen is brushed off the bee's legs to fall into a tray beneath, awaiting collection.

The traps may injure the bee, with the loss of a leg or a wing, and collecting it is said to be stressful to the colony – they are deprived of an important part of their diet – and usually collection is only effected for a few days so as not to cause real harm to the colony.

Pollen in honeycomb stored by the bees for later use.

The pollen in the traps is attractive to a variety of other species, insects, etc. and the pollen may well spoil quickly, typically becoming mouldy. So it's important that the traps are emptied regularly, and steps taken to preserve the pollen – probably freezing or desiccating - and to remove any foreign bodies – bee's anatomy, other insects, and so on, probably by manually picking through it.

Desiccating it will reduce the prospect of mould growing on it, but durability labelling needs to reflect its storage needs – probably hermetically sealed, in a dry place and possibly refrigerated or deep frozen.

In the hive, bees quickly add nectar and this has a preservative effect, and it can be kept in honeycomb until they need it. Even so it will often be found to be mouldy inside the hive.

### *Propolis*

Propolis reducing the entrance to a hive - this would be simple to harvest, by simply cutting it off.

This is the resinous material collected by honey bees and used by them to line the nest cavity – the name means "for the city" and it's used to reduce entrances and seal cracks and crevices.

It has antimicrobial properties, and may be marketed as untreated, or as a tincture in alcohol. Remember that this makes it a compound food, and it may well cross the boundary and become a pharmaceutical. Other presentations include capsules. It may be collected by scraping it off hive parts with a hive tool, or introducing a special plastic mesh that the bees propolise, which is then frozen, which makes the propolis brittle, and the sheet flexed so the propolis breaks off.

The source of the propolis needs to be considered. Bees have been seen collecting from molten pitch – roofing felt and road surfacing – and this potential contamination needs to be considered before deciding to collect and market propolis.[22]

### Royal Jelly

This is the special food that is given by workers to larvae, which changes the development of the larvae into a queen bee, rather than a worker.

It is a specialist crop, requiring special management of the bees to produce, and so harvest in a commercial quantity. The product is rich in protein and pheromones.

It is likely to spoil rapidly, so needs to be harvested and processed quickly, refrigerated, deep-frozen for long term storage.

### Venom

This is harvested mainly for medicinal use, in apitherapy and allergy treatment – not generally as a food.

Harvesting is achieved by exposing the returning foragers to a mild electric shock, which causes them to sting onto a clean glass plate. The venom is allowed to dry, scraped from the plate, and stored in a dark container (it is sensitive to light) in a cool dark place. The process does not harm the bees.

Again, it is protein-rich, and so needs to be processed quickly to prevent spoilage.

## Other products

There are other common products that are made by beekeepers for sale, including honey comb, either as sections or as cut comb, beeswax wraps, and recently I've seen wax "melts" on sale.

### Comb Honey

This includes cut comb, and sections, and the principal hazard is that wax moths may have laid eggs on the comb prior to harvesting. These may hatch after the comb has been removed from the hive, and then the resultant larvae will eat the wax, excreting faeces as they do so.

The controls here are generally to visually inspect the comb quickly after taking it off the hive, then freezing it for 24 to 48 hours. The eggs (which are innocuous) will be killed, and the product is then safe to eat.

If there are any visible faeces on the honeycomb, then it needs to be removed from the food chain, perhaps returned to the bees for them to use the honey.

## Polish

**DANGER** Contains Turpentine
Flammable liquid and vapour; High concentrations of vapour in air may present a vapour/air explosion hazards in the presence of an ignition source. May be fatal if swallowed and enters airways; Harmful in contact with skin or if inhaled; Causes skin irritation; May cause an allergic skin reaction; Causes serious eye irritation; Toxic to aquatic life with long lasting effects: Keep away from open flames and hot surfaces. No smoking. Avoid release to the environment. Wear protective gloves/eye protection. IF SWALLOWED: Immediately call a POISON CENTRE/doctor; Do NOT induce vomiting. IF ON SKIN: Wash with plenty of soap and water. If skin irritation or rash occurs: Get medical advice/attention IF IN EYES: Rinse cautiously with water for several minutes. Remove contact lenses, if present and easy to do. Continue rinsing.

Beeswax polish is often made, and traditional recipes use Gum Spirit of Turpentine as a solvent. This is toxic by ingestion and inhalation, can cause skin irritation, allergies and eye irritation.[23] If it is included as an ingredient in polish, then an appropriate warning is required.

## Beeswax wraps

Many beekeepers are now producing beeswax wraps, which have become popular recently for environmental reasons, as they provide an alternative to plastic containers or a plastic film.

Of course, then the wraps are food contact materials, and so need to be made with food-safe ingredients. The beeswax, any oils or other materials that are added to improve the wrap, including the fabric, all need to be food grade. The wax needs to be carefully processed to keep it clean and free from contamination, the fabric free of dyes and other chemicals that could leach into the food. Natural fabrics may well be acceptable (cotton or linen) but cloths containing artificial fibres such as polyester, nylon, are not produced for food use, and will not meet the requirements of the the Materials and Articles in Contact with Food (England) Regulations 2012.[24]

## Candles, beeswax melts & other products

Beeswax candles sell well, and other products such as melts (diffuser products in law) are outside the food safety legislation, but there is consumer protection legislation that applies, either generally or specifically, and which must be complied with if you are selling the products. The principal one is the General Product Safety Regulations 2005.[25]

There are three British Standards for Candles, dealing with Sooting Behaviour, Fire Safety and Safety Labelling.

There's a summary on the Bromley LBC website, and a bullet point summary:

- the product should be traceable, so consumers need to know the name and address of the producer
- consumers need to receive adequate warnings and instructions for use
- consumers are protected from inhaling or touching harmful chemicals
- goods sold are legal, safe and accurately described
- fire risks are minimised
- candles should be supplied with a safety warning
- if a product (candle, melt or other) is intended to release a fragrance (so scented candles), then each blend needs to have a Safety Data Sheet and Classification, Labelling and Packaging label
- products that claim to be "insect repellent" need to be registered with the Health and Safety Executive.

As said, this is a brief bullet point summary of complex legal requirements.

### Cosmetics

Some beekeepers produce cosmetics – lip salves, hand creams and the like, and these come under specific legislation which I am not covering here. There have been some very good articles in e.g. *BeeCraft* and *BBKA News* that deal with the relevant legislation.

> A "cosmetic product" shall mean any substance or mixture intended to be placed in contact with the various external parts of the human body (epidermis, hair system, nails, lips and external genital organs) or with the teeth and the mucous membranes of the oral cavity with a view exclusively or mainly to cleaning them, perfuming them, changing their appearance and/or correcting body odours and/or protecting them or keeping them in good condition.
>
> Article 2 of the UK Cosmetics Regulation (UKCR – Schedule 34 of the Product Safety and Metrology Statutory Instrument) and the EU Cosmetics Regulation (Regulation (EC) No. 1223/2009)

**Medicinal and Pharmaceutical products**

Honey has apparently been used as a medicine for thousands of years; it is mentioned in Sumerian tablets, and by Aristotle amongst others.[27]

However, beware of making a "medicinal claim" which would be advertising, or labelling it using words such as *heal, cure, prevent, restore, avoid* or *fight*. A claim such as "contains a high source of antioxidants and an abundance of natural enzymes and antimicrobial properties to help boost your immune system" must be seen as suspect; it does contain enzymes and antioxidants and does have antimicrobial properties, but are they "high" and "abundant" and, if so, do they have any effect on the immune system?

> I've often been asked to supply local honey as the customer suffers from hay fever, and it is said that honey contains local pollen and so will lessen the symptoms or even effect a cure.
>
> I've always been happy to sell honey in these circumstances, and customers have come back and assured me that it has been highly efficacious.
>
> There's not much in the way of peer-reviewed scientific literature that supports this, and it should not be made as a marketing claim for honey.

Medicinal claims can only be made about products licensed by, for instance, the Medicines and Healthcare Products Regulatory Agency, or the Veterinary Medicines Directorate. I'm not covering this area of legislation in more detail.

# Food Rooms

Any room where food is processed is a *food room*, and there are legal requirements[28] that apply to it.

## Permanent Food Rooms

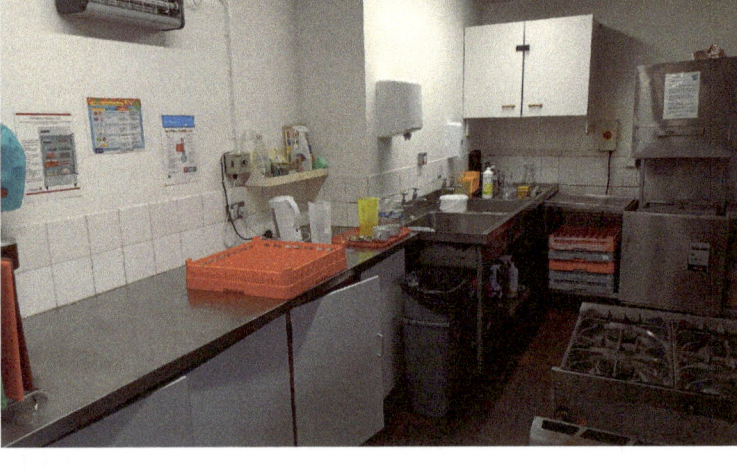

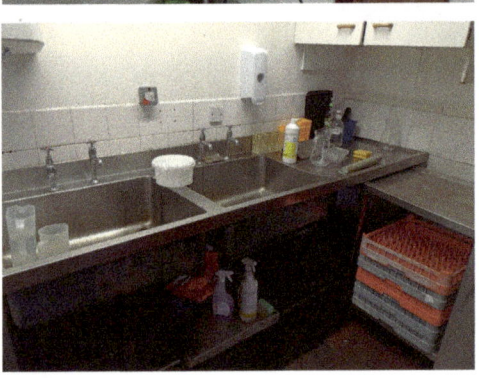

A pair of sinks in a commercial food room - probably much larger than needed by most beekeepers

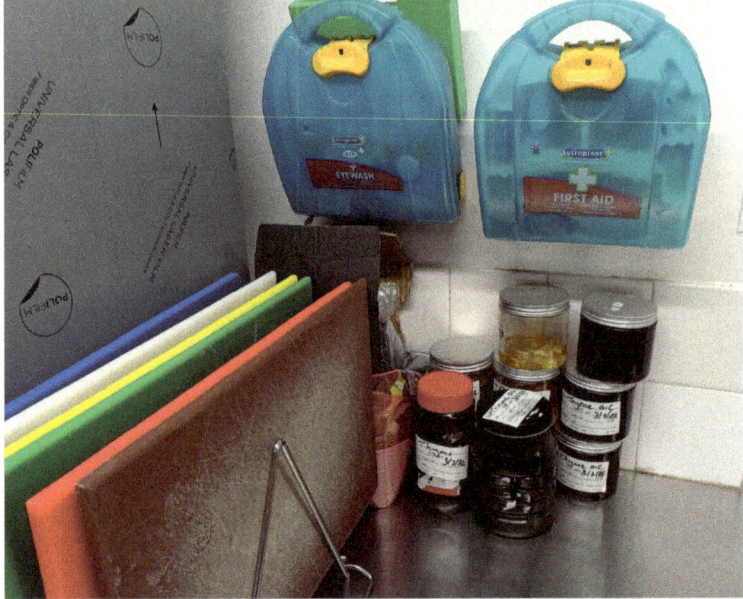

A first aid station in a commercial kitchen; in addition to the first aid kit, you can see ingredients labelled with contents and date, and a set of chopping boards colour coded to help avoid cross contamination. Blue for raw fish, white for dairy and bakery, yellow for cooked meat, green for salad and fruit, red for raw meat, and brown for vegetables

For permanent food rooms ...

- **Floors and Walls** – maintained in a sound condition and easy to clean and disinfect – this requires "the use of impervious, non-absorbent, washable and non-toxic materials".

- **Ceilings and overhead fixtures** (light fittings, pipework, ducts, etc.) – "to be constructed and finished so as to prevent the accumulation of dirt and to reduce condensation, the growth of undesirable mould and the shedding of particles."

- **Windows and other openings** are to be constructed to prevent the accumulation of dirt – fly-screens are required where necessary.

- **doors** are to be easy to clean and, where necessary, to disinfect.

- **surfaces** (including surfaces of equipment) in areas where foods are handled and in particular those in contact with food are to be maintained in a sound condition and be easy to clean and, where necessary, to disinfect. This will require the use of smooth, washable corrosion-resistant and non-toxic materials, unless food business operators can satisfy the competent authority that other materials used are appropriate.

There need to be adequate facilities for cleaning equipment, and these must have hot- and cold-water supplies to them.

There also needs to be a potable (drinking-water quality) water supply. In the UK a direct mains supply should be fine, but beware supplies fed from tanks, and private water supplies need to comply with the Private Water Supplies Regulations 2016.

There need to be adequate facilities for washing food – generally this is seen as being separate from the equipment cleaning facilities. For the beekeeper, it's unlikely that, for instance, vegetables or meat will be being prepared, so it may be acceptable for a single sink to be available.

There also need to be adequate hand washing facilities, and these need to be dedicated for this use. Generally, it's not acceptable to only have one sink in a food room, such is the importance of hand washing – there need to be a sink and a separate "WHB" (Wash Hand Basin).

A range of food safety sings for display in the wc and kitchen

A separate wash hand basin, with liquid soap dispenser above and paper towel dispenser - on the left, a poster showing good hand washing technique.

The toilet must be ventilated, and in such a way that aerosols don't pass from the toilet to the food room and so cause contamination of the food. This means there should be a ventilated lobby between the food room and the toilet.

There needs to be good lighting, both natural and artificial, and good enough to enable safe handling of the food, effective cleaning and the monitoring of cleaning standards.

Outdoor clothes need to be stored outside the food room, as must personal effects.

There needs to be a covered bin and refuse removed quickly.

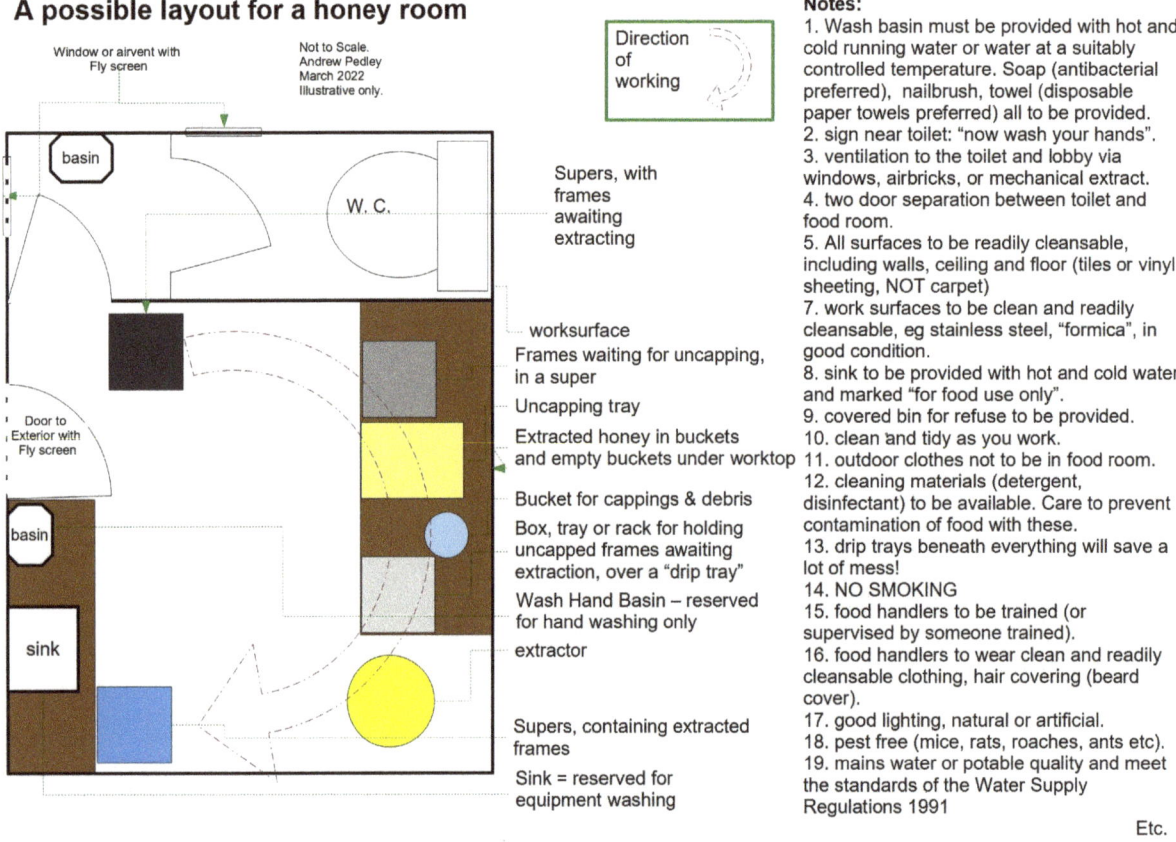

### Requirements for domestic premises

The Regulations accept that some foods are offered for sale having been prepared in a domestic kitchen, and here a different standard is acceptable, – it applies to *"premises used primarily as a private dwelling-house but where foods are regularly prepared for placing on the market"* amongst others, to quote the chapter heading.

The requirements are more pragmatic, and require:

> so far as is reasonably practicable, that they are so sited, designed, constructed and kept clean and maintained in good repair and condition as to avoid the risk of contamination, in particular by animals and pests,

and in particular, where necessary:

> facilities are to be available to maintain adequate personal hygiene, hygienic washing and drying of hands, hygienic sanitary arrangements and changing facilities;

> surfaces in contact with food are to be in a sound condition and be easy to clean and disinfect. This will require the use of smooth, washable, corrosion-resistant and non-toxic materials;

> adequate provision is to be made for the cleaning and, where necessary, disinfecting of working utensils and equipment;

> where foodstuffs are cleaned as part of the food business's operations, adequate provision is to be made for this to be undertaken hygienically;

> an adequate supply of hot and/or cold potable water is to be available;

> adequate arrangements and/or facilities for the hygienic storage and disposal waste are to be available;

> adequate facilities and/or arrangements for maintaining and monitoring suitable food temperature conditions are to be available – honey is of course an "ambient" food and does not spoil at room temperatures. However, the same may not be true of compound foods containing honey, such as bread, cakes or confectionery;

> foodstuffs are to be so placed as to avoid the risk of contamination so far as is reasonably practicable.

Many beekeepers will be using their domestic kitchens, so this section will apply to them.

## *Other requirements for food premises*

### *First Aid*

There needs to be a first aid kit, with waterproof dressings, and any cuts or abrasions should be covered. Note that special "Catering" first aid kits are available with, for instance, special easy-to-see adhesive, waterproof dressings.

Catering adhesive dressings - waterproof and brightly coloured, so easy to see.

### Pests and vermin

Honey is a good food for bees, humans, but also rats, mice, cockroaches, ants, wasps and in all probability a wide range of other creatures.

The Food Safety law requires that food rooms are pest-proof and that there are arrangements for pest control.

Mouse droppings, never a good look in a food room.

Many food businesses will contract established, reputable pest control companies who visit on a regular basis, place baits and traps on a prophylactic basis – so that the errant rat, mouse, etc. will be killed and an infestation be prevented. Part of the contract will be monitoring – regular and thorough checks for signs of infestation – droppings, "rubbing marks" from where animals have rubbed up against walls etc. leaving discoloration, gnaw marks, faeces, eggs, egg cases, larvae can all be indicators of an active infestation, as well as the presence of the actual critters.

The contract will include recording visits in some kind of log, that will be available for inspection, and would be used as part of a "due diligence" defence should a prosecution be undertaken.

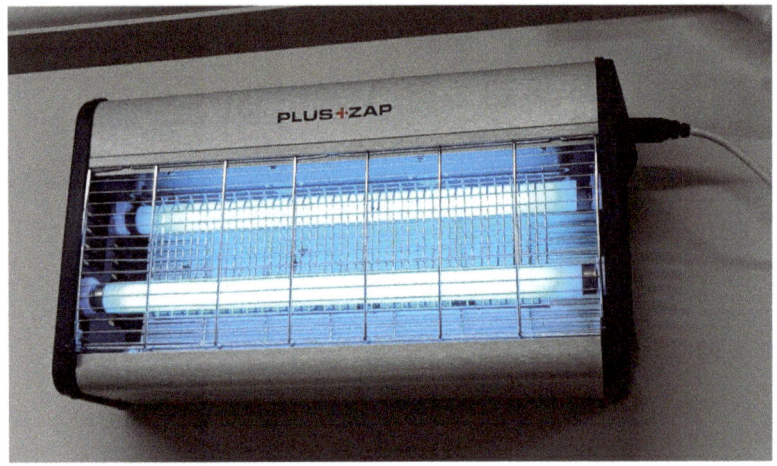

Ultra violet fly killer

Food rooms are likely to be provided with ultra-violet flytraps – insects are attracted to UV light, and fly towards it. In front of the light source will be an electrified mesh, and the insect is killed when it comes into contact with the mesh, its body falling into a catching tray beneath.

It's important that these are sited a little way from the actual food production area; the electrocution of the insect can result in it fragmenting, and bits spreading beyond the catching tray, falling and potentially contaminating food or surfaces beneath.

The Food Safety regulations also require that doors and windows are screened against flies, etc. (Honey rooms need to be bee-tight anyway!)

A kitchen door with a fly screen

# Food contact materials

All the equipment that we use in honey production and packing constitutes food contact materials – the honeycomb, if we pack sections, or cut comb, then the wrapping for those; jars (usually glass but of course plastic jars are available) and their lids (plastic or metal with a plastic insert and a film of lacquer); all our tools – such as uncapping knives, forks, rollers, planes, creaming mixers, etc.; gloves – come into contact with the food; as does the extractor, its cage and the outer casing, the honey valve, ripening tank, honey buckets, strainers – and anything else. They are all food contact materials and could affect the quality of the food, so need to be food grade, and handled appropriately.

The concern is that the food may be adversely affected by the material it is in contact with – chemicals in the Food Contact Material (FCM) may contaminate the food. There are additional provisions for equipment used in food production[29] –

All articles, fittings and equipment with which food comes into contact are to:

| Regulation wording | Discussion |
|---|---|
| (a) be effectively cleaned and, where necessary, disinfected. Cleaning and disinfection are to take place at a frequency sufficient to avoid any risk of contamination; | This requires first that equipment can be cleaned – so equipment needs to be able to be taken apart to be cleaned, as well as being made of materials that are easy to clean. Where equipment needs to be dismantled, the staff must have the necessary skill and equipment (special spanners, etc.) to be able to do so safely. |
| (b) be so constructed, be of such materials and be kept in such good order, repair and condition as to minimise any risk of contamination; | |
| (c) with the exception of non-returnable containers and packaging, be so constructed, be of such materials and be kept in such good order, repair and condition as to enable them to be kept clean and, where necessary, to be disinfected; and | The frequency of cleaning is a matter of judgement, bearing in mind the equipment, the nature of the food and the hazards that it presents – is it necessary to clean your honey extractor after each super (probably not) at the end of each day's extracting, or when the extraction has been finished for a few days? You decide, but be sure you can justify your decision. |
| (d) be installed in such a manner as to allow adequate cleaning of the equipment and the surrounding area. | |

> I have been asked whether plastic foundation could present a hazard, and this would depend on the formulation of the plastic it is made from. If it is "food grade" then the answer should be a No! However, if the plastic is not food grade, then it is possible that chemicals such as Bisphenol A could be present and leach into the honey.

Packaging, equipment, seals on machinery and lubricants that are used on machinery will be Food Contact Materials (FCM), if food comes into contact with them, so they must be suitable for this use.

Many items will be marked with a "cup and fork" symbol – or with the words "for food use", and have been manufactured and are sold specifically for the food use.

Special food-safe greases are available and should be used for honey extractors and other equipment with moving parts, or use a food oil.

> Petroleum jelly has been used in the past as a food grade lubricant – however, investigation suggests that it may contain carcinogenic Mineral Oil Hydrocarbons, and polyaromatics. A product that is labelled as food grade or food-safe would be acceptable, but not otherwise.

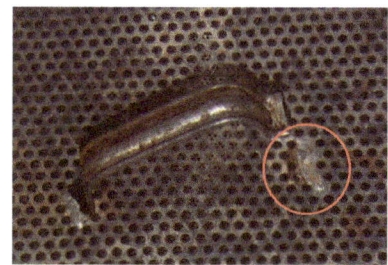

In the case of honey, which is acidic, contact with lead solder in tinplate extractors or ripening tanks can result in the lead-based solder corroding, with the lead dissolving in the honey.

Beekeepers are great innovators and improvisers, and in the beekeeping media, at exhibitions, or on YouTube videos etc., you may well see equipment being made, used, or converted for use in food processing. These should be viewed with some caution – all the components that the food comes into contact with need to comply with the FCM requirements; the device must be capable of being dismantled and cleaned effectively. If it cannot, then it must not be used.

A diy honey press at the National Honey Show - does it comply with regulations?

> Honeycomb itself can be seen as a food contact material. Treating that with any pesticide to prevent wax moth damage can contaminate it and this may lead to the contamination of the honey that the bees will store in it – rendering that honey unfit for human consumption (and potentially harmful to bees, too).

Interactions between food and food contact materials can be complex – and sometimes poorly understood.

# Food handlers

The FSA use the following description in their booklet "Food Handlers: Fitness to Work"–

"... the term 'food handler' mainly refers to people who directly touch open food as part of their work. ... it also includes anyone who may touch food, contact surfaces or other surfaces in rooms where open food is handled. This is because they can also contaminate food by spreading bacteria, for example to surfaces that food will come into contact with, e.g. work tops and food packaging, before it is used. They can also contaminate other surfaces such as door handles, which can then contaminate the hands of people who handle food directly, for example. The term can therefore apply to managers, cleaners, maintenance contractors and inspectors for example. It is the effect of their presence that is important, not the reason for them being there."

The publication is essential reading for anyone running a food business, and there are strict requirements

- Any food handler who suffers from, or carries, any gastroenteritis illness – typical symptoms include diarrhoea or vomiting – must stop food work immediately and not work again for 48 hours after the symptoms have stopped;

- They must wash their hands thoroughly before working, and – very importantly – after using the toilet. This is so important that the FSA have produced a video, which is free to view on YouTube[30] and there's a requirement that you have in each toilet "Now wash your hands" signs. There must also be adequate facilities for hand washing, both in the food room, and in the sanitary accommodation.

- They must "maintain a high degree of personal cleanliness and wear suitable, clean and, where necessary, protective clothing." This may include a clean white coat, hair covering, beard snood. Disposable gloves are not essential, but may be desirable.

- Food Handlers need to have been trained in food safety – there are various courses available, some online, and at the end there is an assessment, and certificates will be awarded. Courses may be accredited, in which case their content, delivery and assessments are monitored externally – and an accredited course should **always** be used. Level 2 is generally considered to be sufficient

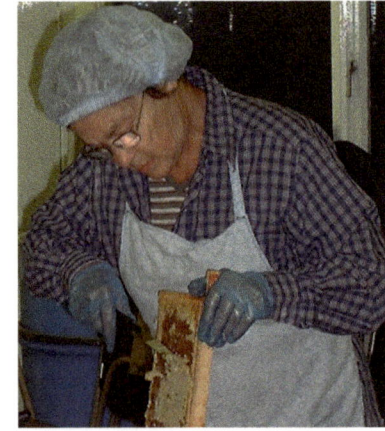

A well prepared food handler; hair net, apron, overclothing, and gloves.

for food handlers, though managers should go for a higher level. Local Authorities will be able to advise on suitable courses.

- Food handlers should avoid hand-mouth and hand-nose contact – hand-washing after e.g. blowing your nose is highly desirable!
- Food Handlers MUST NOT smoke (or vape). It's not just the ash that is of concern, but the hand-to-mouth contact of manipulating the cigarette / cigar/ pipe / vaper.

---

- Many local authorities organise Food Safety Courses, or have arrangements with trainers so that the courses are available to their food businesses; typical cost is around £50 for the day's training and a certificate.
- On completion of the course, a Certificate may be awarded, and these are often displayed in the food business for all to see – good publicity showing the staff and management are aware of food safety. There are some specific courses (Allergen, Labelling, Root cause analysis, Traceability, Vacuum packing) on the FSA website.

---

- Gloves, if used during processing honey, will be food contact materials and so need to be made for food use. Vinyl gloves are manufactured for use in the food business, and biodegradable latex gloves are an environmentally desirable alternative. If suitable, packaging of either type will be marked with a cup and fork or symbol.
- Sometimes there will be exemptions – one pack that I've purchased is annotated "except fatty foods" and another was annotated as not being suitable for use with acid foods, so would not be suitable for use with honey.

# Cleaning

Before progressing with the harvest, preparation is required, and cleaning the equipment and food room where the processing will take place is essential.

This is obviously of great importance for food production, but it can be poorly understood. You do need to have adequate facilities for the cleaning that you are going to undertake – as discussed under Premises, at least one sink is essential, and maybe more, depending on the nature and scale of the work that's being undertaken.

> The 1964 Aberdeen Typhoid outbreak gives an example of cross-contamination that could have been avoided had equipment (in this case, a meat slicer) been effectively cleaned and disinfected.
>
> The subsequent investigation concluded that a tin of corned beef had been contaminated by Salmonella typhi, the causative organism of typhoid fever.
>
> This had been sliced on an industrial meat slicer, which became contaminated with the bacterium, and other meat products were sliced on the same machine, so becoming infected with the bacterium.
>
> As these were cooked meats (ham and the like) they were not cooked prior to consumption, and so the bacteria were not killed, resulting in a major outbreak with 487 hospital admissions.

**Cleaning equipment**
Washing with hot water is essential, with a food-safe detergent or detergent / sterilant (sanitiser) – domestic washing up liquid is good for this. It's essential that all areas are washed down, and scrubbed – adhering dirt and debris such as propolis or wax may need to be removed with an abrasive or a scourer of some kind.

> Some years ago, one of the lectures at the National Honey Show was given by a scientist who subjected honey to microscopic examination; one sample had been presented to him as it "sparked" when placed on toast in a microwave. His examination revealed fine steel particles, probably from a metallic scourer that had been used on the equipment. His suggestion to avoid this was to use a plastic scourer. Later you'll see that one producer had to withdraw batches of honey as they contained "fine metal particles", perhaps from the use of metal scourers.
>
> With the passage of time, the presence of microplastics in the human body is now a concern, and plastic scourers are likely to be a source of these, which are also undesirable.

Once washed, rinse and then dry thoroughly. The rinse can just be clean water, or you may want to soak items in a food-safe sterilant – though this is not necessary. Products sold for sterilising baby's feeding bottles would be appropriate,

Note too that there is no need to sterilise the jars – there's often confusion between honey and jam, where sterilising the jars is normal practice and essential. Thanks to the antimicrobial properties of honey, well-washed jars and lids are acceptable. Nothing else in the process is sterile!

Draining, then air drying, is better than wiping with cloths, which are likely to become contaminated and sooner or later will spread bacteria, etc.

It is important to wash all jars and lids and other packaging – they may appear clean, but jars are likely to have been packaged in the factory by automated equipment, and there is no assurance that they will be clean to food-safe standards. Some manufacturers print a warning on their boxes that the jars should be washed before use. Lids may well have been handled and repackaged by hand, and again there's no assurance that they will be clean to food-safety standards. The items could have been poorly stored before delivery by the supplier or the delivery agent and could have become contaminated. Remember you are responsible, not the supplier.

There is conflicting advice about washing lids to jars – one supplier advises that when lids are supplied fitted to the jars, they do not need further washing. However, one lid manufacturer gives advice on both washing and drying – specifically as when washed, water will be introduced to the "curl" in the rim of the metal lid, and care will need to be taken for this to be dried out.

Here ink has been used to demonstrate how a droplet forms on the punt in a jar

After washing, jars need to be left on their sides for a while, then inverted, and stood on a rack of some kind. This is because the bottom of the jar is usually concave (the "punt"), and water will form a droplet in the centre of the jar – not only will this take a while to dry as the surface / volume ratio is small, it may well leave an annoying mark that mars the appearance of the product.

> As a student, I was asked to go and check on the cleaning that was being used at a local ice-cream parlour; there had been repeated failure of "Methylene Blue" tests indicating bacterial contamination and poor practice.
>
> I attended and watched the operative strip the machine down and wash it before reassembling it.
>
> There were a number of issues that I identified: first the work was being done in a small room with inadequate space to get properly organised; there were inadequate work surfaces; the sink was small, and the hot water supply was inadequate. Dilute bleach was being used, rather than a detergent, so it was not dissolving the food residue. A wiping cloth was in use, rather than a brush. The lighting was poor, so it was difficult to see that the (vanilla ice cream) residue remained on the light plastic components.

While air-drying, the relative humidity in the room will increase and this will both slow the rate of drying, and at the same time may result in the water content of the honey increasing, as honey is hygroscopic. It's important that the area is well ventilated and any honey must be hermetically sealed.

Cleaning up afterwards is important too – the first wash of the actual extractor after an extracting session may be with cold or warm water, as there are likely to be bits of wax in the extractor. You can do without these being melted onto the surface of the extractor – tricky to remove if melted on by hot water, but then it, and all the other equipment used, need to be cleaned up before being stored for next year.

**Cleaning premises**
The premises need to be clean too – and that's all areas that are involved with the food production, including the toilet, communicating doors, and the food room itself. Start at the top, and make sure there are no cobwebs adhering to light fittings, that the walls are clean (wash them down if possible) that the floor is clean – mopping it with a suitable detergent.

Again, allow everything to dry before working with the honey – as honey is hygroscopic and will absorb moisture from the air if the RH is high. It's said that with an RH of 60% the water content of the honey will stabilise at about 18.6%, and at 50% it will be about 15.9%. However, with an RH of 80% the water content will stabilise at a highly undesirable 33.1%. Work surfaces etc. need to be clean too – a wipe-down with a sanitiser of some kind is good.

Do include doors, and in particular handles and the "Finger Contact Areas".

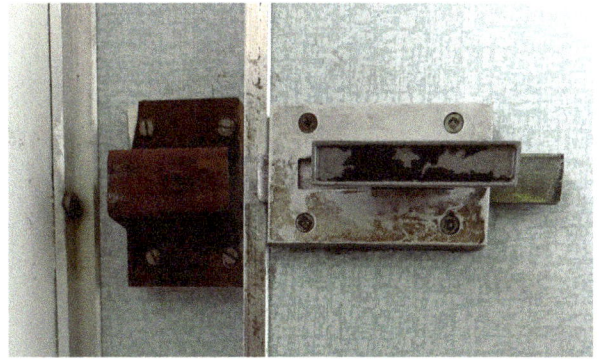

This is the privacy lock on a WC compartment; it is clearly in need of thorough cleaning ... especially bearing in mind the fingers touching the lock will have been engaged in "intimate wiping" a few moments before.

The hand contact area on a door to a busy kitchen; this needs to be completely repainted, the existing paint has been softened by the grease that has collected on it.

## Traceability

This is a fundamental of food safety ... being able to trace where food, and equipment, were sourced. It's one area where documentation is essential.

As a prime producer, the beekeeper will be the beginning of a traceability trail, but the beekeeper will need, even so, to keep records about their honey – different batches being identifiable and documented.

This may be useful for husbandry reasons, or selection of breeding stock ("Hive 3 produced loads of honey, great bees, use that queen as a breeder next year" or "Hive 2 produced little honey, re-queen next year") – but also for food safety reasons.

One reason for batching foodstuffs generally is so that if there is a problem, the defective batch can be identified and dealt with – and you'll often see food "recalls" where a product has been found to be defective and the entire batch recalled. This does seem unlikely with honey, but has been known: in 2016 Hilltop Honey had to recall several batches of honey "due to the possible presence of small pieces of metal."[32]

> **Sudan 1**
>
> Sudan 1 is a dye usually used in shoe polish. It is thought to be carcinogenic, and so it is illegal to use it as a food dye in the UK.
>
> In 2005, a batch of chilli powder was imported that had been dyed with Sudan 1 and been used in food products.
>
> As a consequence, a Food Alert was issued by the Food Standards Agency and all products that did contain, or which could have contained, the contaminated chilli powder were withdrawn. This was facilitated by the traceability and batch numbers.

Some suppliers print labels with sequential batch numbers – so in effect each jar is labelled as a separate batch. To be meaningful and purposeful, this would require records being kept of which numbers related to which batch of honey.

Batch numbers need not be printed on the label: an indelible marking on the jar, clearly handwritten, could be sufficient.

### *Tracing equipment*
Equipment should be traceable too – again, against there being problems in the future. So keeping a record of where each item was purchased, when, and a copy of the receipt / invoice, plus any other documentation would be the ideal.

Ideally, you should be able to trace each item back to its supplier – for a honey farmer, this may be simple as equipment is likely to be ordered and there will be a paper trail from orders, invoices and receipts.

I don't have a comprehensive collection of receipts that show, for instance, that I bought this uncapping knife from XYZ Bee Supplies at the Spring Convention in 2019, which is the ideal of traceability. If there were to be a problem caused by a particular item of equipment, it would be useful to be able to identify its source, and manufacturer.

### *Retail Sales*
You don't need to keep records of retail sales – so if you sell direct from your doorstep, or at a market stall, etc., you don't need to keep notes of every customer who you supply.

### *Bulk sales / wholesale*
However, if you sell in bulk – either in the bucket, or bottled up to a shop or other retailer, then you need to keep a record of that transaction, ideally to batch number level. This could simply be done on the receipt or invoice: "Honey, 20 jars batch number 123xyz" or "Honey, 3 buckets, approximate gross weight 12kg each, batches 123xyz, 124xyz, 125xyz".

This helps to protect you against fraud – a recent Facebook enquiry was from a beekeeper who sold honey, by the bucketful, to his local pub, who then packaged it for retail sale and sold it.

No doubt here there was a friendly relationship between the beekeeper and his totally trustworthy local, but imagine the situation where someone, let's call him Mr Lying-Toerag, purchased three buckets from you, and unknown to you, another ten of syrup or lower quality honey which they blend with yours, then selling it as "Lying-Toerag Honey". Suspicions are aroused, a complaint is made, as in the case of the Bakers, and an investigation launched. The Lying-Toerags, to protect their backs, pass the buck to you … "I got all my honey from Mrs Honest Bee-keeper over the road …" and you are implicated … but you can say, and support with documentation, that you supplied just three specific batches to the Lying-Toerags, 36kg in all, and you have a defence. The buck goes back to them.

It may seem far-fetched, but the price of peace is eternal vigilance, and I've known several instances where honey has been sold by a third party in irregular circumstances.

## Allergens

This ought not concern anyone who is just producing honey, but if you produce compound foods, then you need to be aware.

> A study in the British Medical Journal analysed 101,891 hospital admissions for anaphylaxis. 30.1% were identified as food-triggered.
>
> Between 1998 and 2018 admissions rose from 1.28 to 4.04 per 100,000 per annum.
>
> Deaths, however did fall slightly.

The law has recently been tightened on this, following the death of Natasha Ednan-Laperouse – Natasha's father bought a baguette from Pret a Manger in Heathrow Airport. The baguette contained sesame seeds, to which Natasha had a severe allergy.

At the time, allergen labelling had not been required on the product, as it was "pre-prepared" on the premises, but there would have been a sign advising to ask about allergies in products on display in the shop.

It is a very serious concern – a study[33] showed that hospital admissions for allergies had increased threefold from 1998 to 2018, though fortunately fatalities have decreased – but there are still fatalities. Between 1 and 10% of the population have a specific food allergy.[34]

Following campaigning by Natasha's parents, the law now requires **all** foods to be labelled appropriately.

There are 14 specified allergens, as follows:

- celery,
- cereals containing gluten (such as barley and oats),
- crustaceans (such as prawns, crabs and lobsters),
- eggs,
- fish,
- lupin,
- milk,
- molluscs (such as mussels and oysters),
- mustard,
- peanuts,
- sesame,
- soybeans,
- sulphur dioxide and sulphites (if they are at a concentration of more than ten parts per million) and
- tree nuts (such as almonds, hazelnuts, walnuts, Brazil nuts, cashews, pecans, pistachios and macadamia nuts).

The legislation also requires that the allergens be **emphasised** in some way in the list of ingredients– **bold font**, larger lettering, underscoring, colour, or *italics* for instance. The legislation also requires that the ingredients list is not obscured in any way, even by a crease.

Keeping track of allergens is essential in compound foods; here a simple label does the job (and records the contents and the date it was made for stock rotation and control benefits)

If an ingredient is a "compound ingredient" – such as mayonnaise – then the ingredients must be listed immediately after the name, so, for instance "mayonnaise (**eggs**, oil, water, salt)".

There is a lot of media attention on this – incidents are reported; tragically there are typically several deaths each year. Although the numbers have been reducing in recent years the number of food-related admissions to hospital has increased.

Breaches are taken seriously in Court – Poundstretcher were fined £24,000, with £6,950 costs and £181 for the absence of allergen labelling, and had to withdraw products from hundreds of their stores. In April 2021, Mohammed Nazamul Islam was fined £450, required to pay £1,000 costs, and a victim surcharge of £45. A product had been purchased from his shop, and found to contain milk, an allergen, despite having been told in the telephone order that the consumer had a milk allergy.

## Second-hand equipment and recycling jars

There's always been a lot of interest in, for instance, reusing honey jars and repurposing other jars for use with honey. It is environmentally, and economically, desirable and reusing milk bottles has been practised in the UK by the dairy industry for many years, collecting, cleaning and then reusing milk bottles via a doorstep delivery and collection service – it been estimated that most bottles were reused 18 times during their lives, with some being 2 years old and having been reused 60 times.

Historically others have recycled bottles – in my lifetime, soft drink bottles had a deposit (4d – old pence), which was refunded when they were returned. Today, in Germany, much food packaging is reused, and there's a refund on beer bottles, for instance.

The FSA gave the following advice in September, 2021:

*"Reusing glass containers, like jam jars, occasionally to supply food such as home produced honey does not present a food safety concern, provided that good hygiene practices are met for the food, and food contact materials are used. Where jars are reused, for example at village fetes or other occasional events, they should be free from chips and cracks, and should be clean and sterilised prior to each use.*

*Lids should not be reused, as they have a gasket within that is designed for single use. Films are unlikely to be suitable for honey, so appropriate lids will need to be procured."*

The statement then continues with information applicable to community and charity events.

This statement calls into question other second-hand equipment – many years ago, a friend gave me an uncapping knife – is that acceptable to use? What about a second-hand honey extractor? There is a good market in second-hand equipment – many Beekeeping Associations have auctions, there are often adverts on eBay.

The answer calls, I believe, for judgement, and a number of factors need to be taken into consideration:

- **Condition:**
  If the equipment is in good condition, then it passes the first test. But if it's seen better days, is damaged, and been patched up, then it's a fail. Plastic becomes brittle and cracks, can be abraded – scratches are hard to clean.

  Even if the item were food grade originally, repairs are likely to have compromised this – adhesives and paints are not generally food grade, and repairs may leave hard to clean cracks and crevices.

- **Construction:**
  Is the equipment fit for purpose by today's standards?

Many extractors and settling tanks on the second-hand market were manufactured in the 1930s or even the 1920s. Common materials then were tinplate, with bronze valves, soldered together with lead-based solder. The solder here is a real issue – acidic honey will dissolve the lead, causing lead contamination.

They would not pass as food grade today, and so must not be used.

Upcycling, for instance by painting rusty components, is very unlikely to produce an acceptable result, and the paint will become a food contact material so needs to be compliant with the relevant regulations.

On the other hand, if it is constructed from stainless steel, or food grade plastic, then they may well pass the second test.

▶ **Manufacturer's intention:**
This may sound a strange one, but was the equipment made for use in commercial food (ideally specifically honey) production, and was it intended to be used repeatedly or only once?

Many items will be durable and expected to last for many uses, so will be fine.

▶ **History**:
In an earlier advice about re-using packaging, the FSA mentioned one reason why milk bottles could be re-used was that there was a "closed loop distribution system" … in other words, the bottle went out from the dairy to the consumer, then back to the consumer. It would be rare for a bottle to go back to a different dairy from the one that supplied it, and often the company name would be moulded into the glass. The dairy, therefore knew the bottle's history, that it was made to be reused, and had only been out to their customers. There was a degree of traceability, although strictly there is a weakness in the traceability chain as the dairy cannot know which customers a bottle has gone to, or whether the bottle has been put to a non-food use by a customer before being returned. But this does not seem to be significant with the potential hazards being managed by thorough inspection and cleaning.

This jar is made from recycled glass, with a cork lid. Though attractive, unless it was made for use with food it should not be used, and the cork lid does not hermetically seal so moisture will be absorbed by the honey, which will ferment.

▸ **Provenance**:
This is related – how confident are you that the previous owner treated the equipment properly and well, so that it will not prejudice the safety of your produce should you acquire it? It's obviously better to get it from someone you know than an unknown vendor on eBay, though the eBay Feedback system may give some indication.

**Summary**:
I've used milk bottles as an example: however, it seems to me that the FSA's 2021 advice effectively **prohibits** the re-use of honey jars for honey for sale.

I hope that this provides useful guidance on the principles that apply when purchasing and using second hand equipment.

Personally, I use quite a lot of pre-owned equipment; however, it is all of good quality, food-safe materials. Generally, most was made for use in honey production. Some items I have made myself – a wooden "uncapping bridge" and some items from our domestic kitchen get used for honey production – colanders and a stainless-steel roasting tray.

## The Honey Regulations

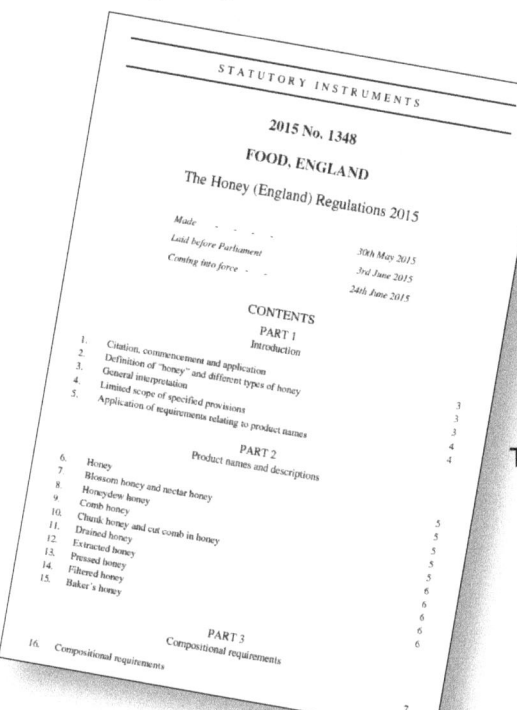

The full title is "The Honey (England) Regulations 2015", though similar regulations will apply in Northern Ireland, Scotland and Wales, thanks to the harmonisation of legislation under the EU prior to Brexit.

There are a number of Sections and Schedules, and I'll do my best to summarise them here – but do please see the regulations (freely available online),[35] to make your mind up for yourselves.

The first section deals with definitions of Honey, as follows:

## Definitions of different honeys

| Regulation | Comment |
|---|---|
| "honey" means the natural sweet substance produced by Apis mellifera bees from the nectar of plants or from secretions of living parts of plants or excretions of plant-sucking insects on the living parts of plants which the bees collect, transform by combining with specific substances of their own, deposit, dehydrate, store and leave in honeycombs to ripen and mature | This is the legal definition of honey in the UK, and note that it includes honeydew, but excludes the product of any species of bee other than Apis mellifera, the European honey bee. |
| "baker's honey" means honey that is suitable for industrial use or as an ingredient in another foodstuff which is then processed | In effect lower-quality honey; there are other specific requirements summarised below. |
| "blossom honey" and "nectar honey" mean honeys obtained from the nectar of plants | This will be the majority of honey that is on sale in the shops, but see honeydew below. |
| "chunk honey" and "cut comb in honey" mean honeys which contain one or more pieces of comb honey | An attractive presentation of honey for retail sale, where a piece of honey comb is inserted into the jar which is then filled with liquid honey |
| "comb honey" means honey stored by bees in the cells of freshly built broodless combs or thin comb foundation sheets made solely of beeswax and sold in sealed whole combs or sections of such combs | Note there is a requirement for "thin" foundation sheets when producing this, or you could use completely natural comb from the use of starter strips. |

| | | |
|---|---|---|
| "drained honey" means honey obtained by draining de-capped broodless combs | This is a method of separating honey and the comb which may be used with, for instance, top bar hives where there are no frames so the comb cannot be centrifuged. The combs are uncapped, but not crushed or pressed: see "pressed honey" below. I have heard it said that draining gives a superior product, as the aromatics that give flavour are not lost in the way that they with centrifugal extraction. | <br>© Diana Ivanova \| Dreamstime.com |
| "extracted honey" means honey obtained by centrifuging de-capped broodless combs | The majority of honey sold in the UK will have been extracted by a centrifugal process. | |
| "filtered honey" means honey obtained by removing foreign inorganic or organic matters in such a way as to result in the significant removal of pollen | The general advice is that "filtering" should not take place, only straining. Given that the majority of pollen is less than 100 microns, it would require commercial equipment to filter the honey so as to remove a significant amount of pollen, as noted above. | |
| "honeydew honey" means honey obtained mainly from excretions of plant sucking insects (Hemiptera) on the living part of plants or secretions of living parts of plants | This would be honey that is derived from aphids – it is a rarity in the UK though it is imported and available for retail sale. |  |
| "pressed honey" means honey obtained by pressing broodless combs with or without the application of moderate heat not exceeding 45° Celsius. | Pressing is an alternative process to draining and centrifuging, and may well be used in the production of Heather (Ling) honey, which is thixotropic, so cannot be centrifuged easily. |  |

… and there is then "General interpretation", much of which is beyond the scope of this book, but some specifics are worth mentioning:

"ingredient" means any substance or product, including flavourings, food additives and food enzymes, and any constituent of a compound ingredient, used in the manufacture or preparation of a food and still present in the finished product, even if in an altered form; residues shall not be considered as "ingredient"';

Label – refer to the Food Information for Consumers regulation which gives the following definition:

> "label" means any tag, brand, mark, pictorial or other descriptive matter, written, printed, stencilled, marked, embossed or impressed on, or attached to the packaging or container of food;

Part 2 deals with product names and descriptions, and either just the word "honey" or one of the descriptions above must be used where it is applicable.

**Filtered honey and Baker's Honey**
There are some specific controls on both "Filtered Honey" and "Baker's Honey", including that the full description ("Filtered Honey" or "Baker's Honey") **must** be used – just the word "honey" is not acceptable in this context.

In order to determine a batch of honey's water content, to determine whether it needs to be labelled as Baker's Honey, you will need to use a refractometer.

There is a very detailed table showing the "presumed" refractive index at 20°C for water contents from 13% to 25% in the Codex Alimentarius.

A refractometer is a simple device: a drop of honey is put onto its glass plate and a transparent cover lowered onto it. The device is held up to the light, and a reading taken by looking at the scale through an eyepiece.

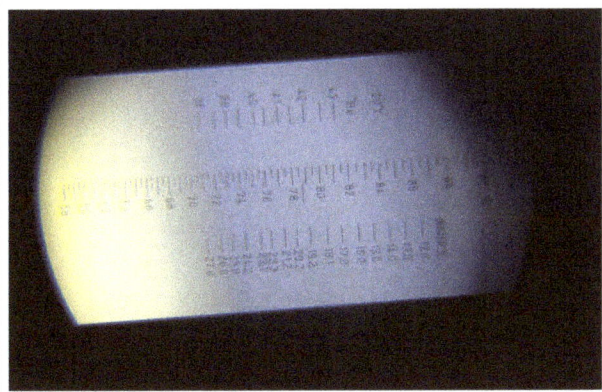
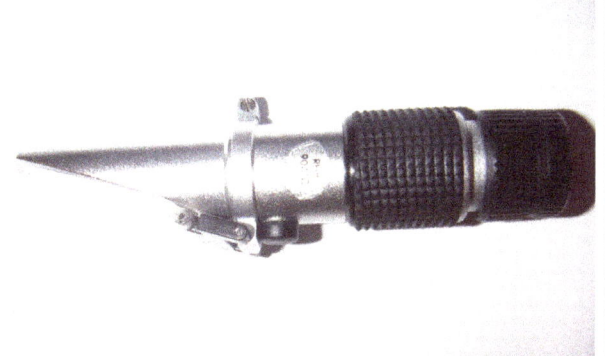

A refractometer and it's scale

These can be affected by temperature: some are marked "ATC" which means Automatic Temperature Compensation.

Refractometers are used for many purposes, measuring salinity in seawater, alcohol in wine, contaminants in engine exhaust condensate, as well as sugars, and it's important to get one with the right scale (Brix) and calibration.

Sometimes refractometers are supplied with calibration liquid, or a calibration plate; if not, then the liquid can be purchased separately.

Calibration using, Extra Virgin Olive Oil, is referenced in the beekeeping literature and forums but the accuracy of these would be open to question if it was relied on as a defence.

Neither filtered honey nor Baker's Honey can have any indication of the "floral, vegetable, regional, territorial or topographical origin of the product or specific quality criteria for the product".

Baker's Honey must have the words "intended for cooking only" on the label close to the product name.

If Baker's Honey is used as an ingredient in a compound food, the word "honey" can be used in the description, but the phrase "Baker's Honey" would be used in the list of ingredients. So, for instance, bread or cakes made with honey as an ingredient could be labelled "Honey Bread" or "Honey Cake", though in the list of ingredients the full "Baker's Honey" description must be used.

## Country of Origin

The Honey Regulations require the country where it was harvested to be specified – the use of the country to be part of the description "English Honey" or "A B Lane, Apiaryshire, England". This does not apply to Filtered or Baker's Honeys, when used as an ingredient. Since 1 Jan 2021, the United Kingdom is no longer part of the EU, so labelling on blended honey would need to reflect that. The FSA website[36] has detailed information – the situation is particularly complex in the case of Northern Ireland, thanks to the Northern Ireland Protocol protecting the Good Friday Agreement.

## Floral Sources

Specific claims about the floral source (Lavender Honey, etc.) are restricted by Honey Regulation 17(3) which states *"the product comes wholly or mainly from the indicated source and possesses the organoleptic, physico-chemical and microscopic characteristics of the source."* So before labelling something as being from a particular flower or crop, you need to be confident that the honey is "wholly or mainly" from that flower type, and the smell and taste must be consistent, as must the viscosity, electrical conductivity, analysis, and pollen content of the flower type.

Any images on the label must not be misleading – for instance, a picture of a bee foraging on lavender could be interpreted that the honey is Lavender Honey, so should be avoided.

Ling Heather Honey at the National Honey Show

While there is no legal definition of, for instance, Heather Honey, you can use the description if the harvest was "wholly or mainly" from Heather, and has the characteristics of Heather Honey.

Table 5.2 in Eva Crane's wonderful *Honey, A Comprehensive Survey*[37] book gives some of the expected characteristics of many types of honey, and the Honey Regulations themselves contain (Schedule 1) "compositional criteria" for some types of honey.

## Geographical origin

Regulation 17(4) of the Honey Regulations allows that the product name may be supplemented by information about a regional, territorial or topographical origin so long as "product comes entirely from the indicated origin." So, for instance, "Oxfordshire Honey" or "Chalfont Honey" could be acceptable so long as the honey comes entirely from Oxfordshire or the Chalfonts respectively. There are grey areas here: what if the hives are on the Oxfordshire / Buckinghamshire border? … Foraging may have taken place in Buckinghamshire as well as Oxfordshire. If the hives are in the centre of Oxford, then "Oxford City" honey may be an acceptable description, but what if they are in Hampton Poyle, a hamlet to the North of Oxford? – the bees are bound to forage beyond the hamlet's boundaries. So labelling the honey as "Hampton Poyle Honey" could be problematic – but only the Courts can interpret the law.

> An example of Honey Fraud that made it to the Courts was the case of William and Lynn Baker, who were convicted of 12 charges "obtaining property by deception" at Norfolk Crown Court in December 2005.
>
> It was found that they had been selling honey from China and Argentina as "Norfolk Honey": a mere 17.7 tons in total. The fines were £5,000 and £3,000, respectively.
>
> Costs were awarded, at £35,000 each, plus they had to pay their own costs of about £10,000, making a total of about £90,000.

Of course, there are analytical techniques that can be used to identify the source of honey, with considerable accuracy – pollen analysis, which is within the scope of beekeeper with a little training and the right equipment. DNA Metabarcoding, NMR and the like are all available, at a price.

Honey at the National Honey Show clear and granulated, dark and light, but it's all honey

**Enzymes**

The Honey Regulations require that the honey must not "have been heated in such a way that the natural enzymes have been either destroyed or significantly inactivated".

The effects of heat on an enzyme are commonly measured by the time it takes to reduce half of the enzyme's activity or its "half-life" at a given temperature. For instance, the half-life of diastase in honey is 1,000 days at 20°C, 14 days at 50°C and 30 seconds at 80°C. The other enzymes in honey are affected similarly. It does not return when destroyed by heat.

Two interesting side notes are that almost all the enzymes in honey are introduced by the bees, and apparently all break down when liquefying crystallized honey in a microwave.[38]

**Other Compositional Criteria**

There are "Compositional Criteria" in the first Schedule, which I'll reproduce in full in Appendix 2, but in summary:

- (it) will contain sugars and other substances, enzymes and solid particles from honey collection
- fluid, viscous or crystallised
- flavour and aroma vary but will be from plant origin
- no ingredient will be added, including any food additive
- no other additions other than other honey (blending)
- free from organic and inorganic matters "foreign to its composition", as far as possible
- must not have any foreign tastes or flavours, be fermenting (permitted in Baker's Honey), have artificially changed acidity, or been heated in such a way as to affect the natural enzymes
- must not have any pollen or constituent particular to the honey removed unless unavoidable (does not apply to filtered honey)

There's then tabulated information about the anticipated characteristics of the specific classes of honey described, and it's here that the limits on water insoluble content, electrical conductivity, free acid, Diastase activity and hydroxymethylfurfural (HMF) content are defined.

None of these can the beekeeper affect, except by some mistreatment of their honey crop, nor can they determine whether their honey complies or not without laboratory analysis.

There are specialist laboratories that offer honey analysis, and in many cases the beekeeper will have little or no control over the values – however each is a measure of the honey's quality or purity, so it is important to work carefully so as not to affect these, and so inadvertently increase the levels.

Dealing with each in turn,

| Water insoluble content | This is used as a criterion of the honey's cleanliness – high levels indicate that the honey processing has been poor. |
|---|---|
| Electrical Conductivity. | This is measured in microSiemens, and is a good measure of the botanical origin of the honey, and may be used in routine honey control. |
| Free Acid | The acidity of honey is important for taste, indicates the freshness of the honey, and indicates that unwanted fermentation has not taken place. |
| Diastase activity | This is a quality indicator, relating to honey storage and temperature. There is a large natural variation. |
| Hydroxymethylfurfural (HMF) | This is a breakdown product of fructose, one of the main sugars present in honey, in the presence of acid (a Maillard Reaction). It is not toxic, but it is an indicator of the quality, the age and the heating that the honey has been subjected to.<br><br>There are various reasons for warming, or heating honey, for instance:<br>▸ to re-liquefy crystallised honey, or<br>▸ as part of the process to make creamed / soft set honey, or<br>▸ to pasteurise it (killing off yeasts so prolonging its shelf life), or<br><br>to artificially evaporate excessive moisture and so artificially "ripen" it |

HMF is of course important, but in fact research indicates that slight heating over a short period of time has a negligible effect on the HMF levels – there are several sources that I've found, including in *Honey, A Comprehensive Survey*, Table 5.84/1 and there is consistency with *Effect of Storage and Processing Temperatures on Honey Quality,* which found:[39]

Time for 30mg/kg HMF to accumulate (based on 3 samples)

| | |
|---|---|
| 30°C | 100-300 days |
| 40°C | 20-50 days |
| 50°C | 4-10 days |
| 60°C | 1-2.5 days |
| 70°C | 3.5 days |
| 80°C | < 2 hours |

A "DIY" method of determining HMF was published in *BBKA News* No. 162, 2006.[39] The accuracy of this is not known by me, and it should not be relied on for legal purposes.

It must be borne in mind though that the HMF limit would be measured at the time of sale, so minimising the time between harvest and sale is desirable, and storing the honey prior to sale somewhere cool is also desirable. The BBKA guidance used to suggest storage at 10°C if possible.

**Specific quality criteria**

The Honey Regulations allow "The product name of a relevant honey may be supplemented by information relating to its specific quality criteria", and it is this that could legitimise descriptions such as "raw".

The only legal definition for "raw" honey that I've been able to trace is in Utah, USA,[40] where:
> *"Raw honey" means honey:*
> *(i) as it exists in the beehive or as obtained by extraction, settling, or straining;*
> *(ii) that is minimally processed; and*
> *(iii) that is not pasteurized.*

Dealing with each of those in turn,

- *obtained by extraction, settling or straining* these are the common processes for honey, this seems to exclude pressing and draining though;
- *that is minimally processed* beyond straining: most honey is unprocessed – except perhaps creaming;
- *that is not pasteurized.* Pasteurisation temperatures are 63°C for 30 minutes or 77°C momentarily, well below the temperatures that the vast majority of honeys are subjected to.[41]

So Utah's raw honey description includes the vast majority of honey, probably including a lot of mass-marketed "supermarket"-type products.

Other claims are made about honey that is described as raw, such as:

"*Raw honey is only strained before it's bottled, which means it retains most of the beneficial nutrients and antioxidants that it naturally contains. Conversely, regular honey may undergo a variety of processing, which may remove beneficial nutrients like pollen and reduce its level of antioxidants.*"

Of course, the Honey Regulations prohibit filtering so as to remove pollen, so that definition is axiomatic. Removal of antioxidants by filtering is highly unlikely: the molecules would be far too small.

Another definition claims that, "*Some regular honey products contain added sweeteners, such as high fructose corn syrup.*" Again, the Honey Regulations prohibit the addition of any other ingredients to honey, and the addition of other sugars would constitute adulteration, a fraud. Presumably those stooping to adulteration or other fraud would not be averse to labelling the adulterated or counterfeit honey as "raw" either – there is no regulation on the use of the word as there is about "organic" claims.

The Association of Chief Trading Standards Officers have considered the use of "raw" in connection with honey and have produced guidance for their officers. This is not a public document, and may be interpreted differently by different Local Trading Standards Departments, so if in doubt, contact your local Department for advice.

The term "raw" does not seem to be a specific quality criteria, although there is no case law. Enforcement of Regulation 7(1)(c) of the Food Information Regulations for Consumers (EC 119/2011) would probably be by service of an Improvement Notice rather than prosecution. It seems that at least a couple of beekeepers have been told by their Local Authorities to stop using the description, either informally or via a formal Notice.

### Other superlatives
In addition to the word "raw", all of these have been seen in use as descriptions of honey:

> - **Artisan** is presumably intended to show that the producer meets the definition of an Artisan, viz.: "*a worker in a skilled trade, especially one that involves making things by hand*" or "*(of food or drink) made in a traditional or non-mechanized way using high-quality ingredients*" (Oxford Languages) – but who decides who is an "artisan" and entitled to use that description?
>
>   And surely the honey is produced by bees, not Artisans?
>
> - **Pure** – of course, the Regulations prohibit the addition of anything to the Honey, so it is by definition pure, and this description is also axiomatic.

© Elena Ray Microstock Library
© Elena Ray | Dreamstime.com

- **Natural** – again, the Regulations require that honey is the product of the Apis mellifera, the Honey Bee, so it cannot be anything but natural.

- **Active** – apparently relates to antimicrobial activity of the honey, after testing by an independent laboratory; the term does not seem to be in common use, and there does not seem to be a legal definition of it anywhere, though research suggests that it is used in relation to Manuka honey with a UMF of 10 or more. It could be argued this is the only one that could be considered to be a Special Characteristic under the Honey Regulations and so justify the use of the description.

- **Countryside** – which implies that none of the hives are urban, and perhaps that the bees do not forage in conurbations. How would you prove that?

Other superlatives I have seen are *Vibrant, Living, Unblended, "Harvested by loving beekeepers"*, and *"Ethically harvested"*.

While the Honey Regulations do allow specific quality criteria to be used to enhance the description of honey, the EU Regulation (EU) No 1169/2011 of the European Parliament and of the Council, Article 7(1)(c), reads:

> *Food information shall not be misleading, particularly:*
>
> *(c) by suggesting that the food possesses special characteristics when in fact all similar foods possess such characteristics, in particular by specifically emphasising the presence or absence of certain ingredients and/or nutrients.*

And general consensus amongst beekeepers seems to be that these descriptions are axiomatic and superfluous.

## Registration

This is a principal plank in food safety – all food businesses must register with the Local Authority; it is an offence not to do so. The Legislation reads as follows:

> "every food business operator shall notify the appropriate competent authority, …. of each establishment under its control that carries out any of the stages of production, processing and distribution of food, with a view to the registration of each such establishment."

The reason is simple – food businesses are subject to a range of regulations, for public safety, and enforcement is important, so it's important to know where the businesses – and their premises – are for them to be visited and assessed for compliance.

The actual legislation is in **Regulation (EC) No 852/2004 of the European parliament and of the council of 29 April 2004 on the hygiene of foodstuffs** and in essence, it requires ALL food businesses to register, though there is an exemption in the case of prime products, which includes honey, thus beekeepers.

The exemption is as follows:

- The direct supply, by the producer,
- of small quantities of
- primary products to the
- final consumer or to local retail establishments directly supplying the final consumer;

Then there's specific guidance about honey production, which can be seen as primary production, as follows:

> "Honey and other food from bee production: all the beekeeping activities must be considered as primary production. This includes beekeeping (even if this activity extends to having beehives at a distance from the bee-keeper's premises), the collection of honey, its centrifugation and the wrapping and/or packaging at the beekeeper's premises. Other operations outside the beekeeper's premises (e.g. the centrifugation and/or wrapping/packaging of honey) cannot be considered as primary production, including those carried out on behalf of beekeepers by collective establishments (e.g. cooperatives)."

So it seems to me that there are four tests in the decision whether you need to register or not, as follows:

1. **Is the food the product of the beekeeper's own hive(s)?** – honey, beeswax, propolis, royal jelly, all would be. But compound foods – foods with ingredients – such as cakes, biscuits, fudge, bread etc. would not be prime produce, and so registration is required. It seems to me that if the beekeeper buys any honey in bulk, then they fail this test and the premises must be registered.

2. **Is the processing exclusively in the beekeeper's premises?** If so, then the prime produce exemption applies. But if you use an Association's Extracting Room, or pop round to a friend's, then this exemption does not apply. Both the Association's and the friend's premises ought to be registered.

3. **Is it only small quantities being produced?** I've heard several arguments about what is "small", including the produce of five hives, as being the minimum viable number of hives for a stable beekeeping operation, through to – "I'm not registered as I'm not a bee farmer".

   The guidance does give suggestions for other food products – for instance, in the case of milk producers – up to 24 pints of raw drinking milk per day – and egg producers – less than 360 eggs per week. For fishermen, it's up to 25 to tonnes of fish in a calendar year. There is no similar guidance for beekeepers and honey production, so it would be for a Court to decide, and presented with information that you considered a particular quantity of honey to be small, bearing in mind that in a good year a single hive can produce a 40kg surplus and on the five-hive basis that represents 200kg of honey, they may or may not agree.

4. **Are sales direct or through a "local" retailer?** Selling direct – farm gate, market stalls, village fetes, etc. are all fine, exemption applies. Does "local" relate to the apiary or the Food Premises (the beekeeper's home, perhaps)? When I lived in London, I had hives in my daughter's garden in Oxfordshire. The honey was produced in Oxfordshire, processed in London, sold direct but also through the village butchers. Is it local when there's 40-odd miles between the hives and the processing, and between the processing and the retailer. Who knows? Only the Courts could rule, based on legal arguments presented to them.

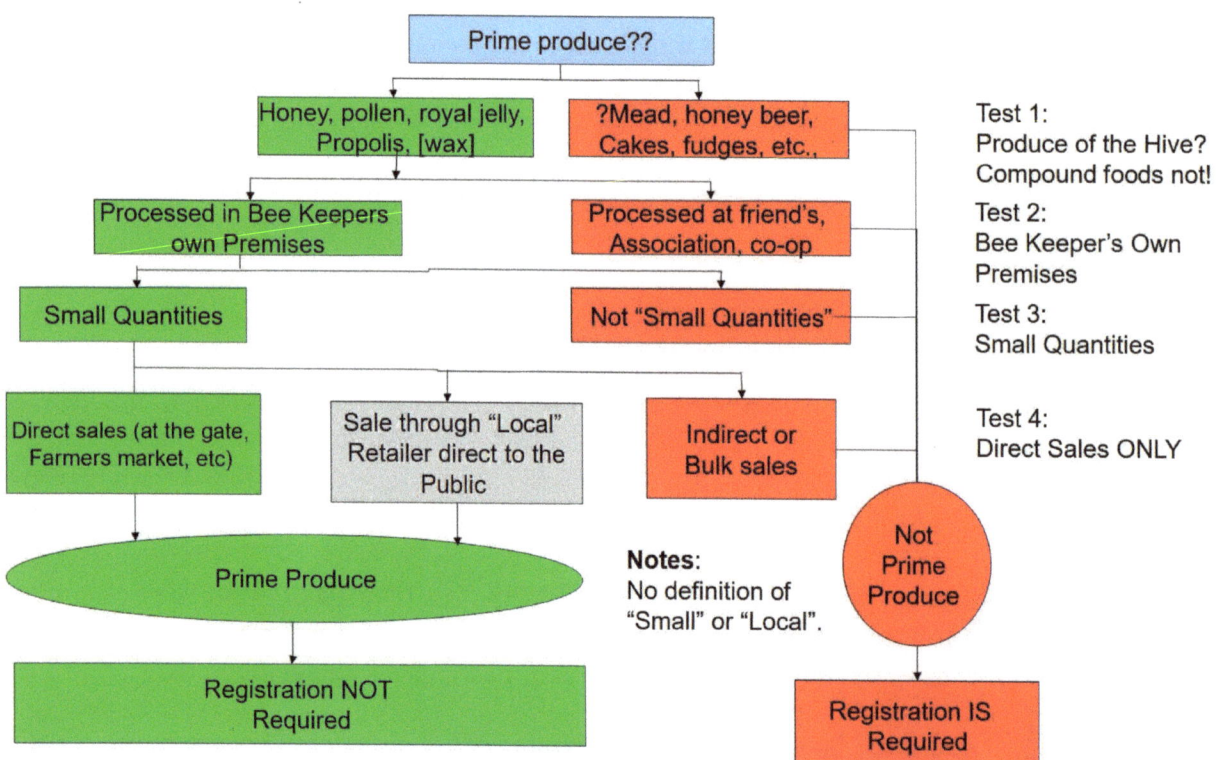

This algorithm may be helpful in making the decision.

I would add though that registration is free, and all the other food safety legislation applies whether you are registered or not. In general the Local Authority Officers will be interested and may well arrange to inspect – their prime role is education so they'll be interested and helpful. They will award a Food Hygiene Rating System (Scores on the Doors) grade, which is good for public relations and marketing.

Registering also future-proofs you. Beekeeping is famous for "creep" – you start with a couple of hives, produce a few jars, but after a couple of years you find you've got 10 hives and sold a lot of honey, moving you further into the grey area of "small-scale".

A change in your business model may also lead to a need to register – for instance, perhaps a restaurant starts to buy your honey in the bucket. This is not "local retail sale" and so the registration requirement applies.

There may well be public liability insurance implications too: check with your insurer.

It also prevents surprises – the local authority may purchase a jar of honey as part of a sampling programme, or notice honey on the shelves when undertaking an inspection of a shop where you sell, and pick up your name and address from the label on a honey jar. And, of course a disgruntled customer may make a complaint about your produce: if you are registered, then you are on the Local Authority's radar … they know who you are, and how to contact you. But if there's no registration, then they are likely to call in person, making enquiries both about the complaint but also about a possible offence of having failed to register.

A good rating is something to be proud of, an incentive to customers who can be confident of your products.

Registration needs to be applied for 28 days before the food business commences, and the offence is absolute. However, if you've been selling produce and are not registered, then it is much better to apply late than not at all!

In my last local authority, I was inspected and got a 5-star rating. Having moved, I applied and am "awaiting inspection". Inspections are undertaken on a risk-based priority basis, and it seems I am considered low risk / low priority.

The registration process is straightforward – there's a simple form to fill in and send off to your local authority – you'll usually find this on their website. This will likely be followed up by a letter, which acknowledges the application, and includes a form asking for a lot more information – this can be quite intimidating, and over the top for the small-scale beekeeper with a few hives putting a little honey out on retail sale. But it's an information-gathering process, and the same form is sent to everyone who applies to be registered, regardless of the size of the operation. Complete it as best you can – phone or call in at the Council's offices if you have doubts or concerns, and they'll assist you, and you can return the form with a covering letter or email explaining your business in more detail if you think that would be helpful.

You should keep the local authority informed if there are significant changes that affect the registration – changes in the business model, such as buying – or selling – honey by the bucket rather than in jars, or starting to produce "value added" products instead of just honey, and also if you cease trading.

## Food crime & Fraud

There are several types of food crime, and for the beekeeper, a principal one is honey fraud.

There's a huge concern at the moment, with honey being one of the most "forged" foods in the world. This is having economic consequences on beekeepers, and not only honey production, but also crop pollination, as it reduces the quality of crops and so the yield to farmers. Other environmental impacts will follow too.

In 2019, the global Apimondia Exhibition disqualified 45 per cent of all entries which were found to be adulterated. The organisers did not reveal the reasons for rejection but could include "such things as illicit sugars, antibiotic and pesticide residues, HMF and country of origin discrepancies". As a consequence, they produced the Apimondia Statement on Honey Fraud V.2 and their definition of fraud can be summarised as follows:

> Honey fraud is a criminal and intentional act committed to obtain an economic gain by selling a product that is not up to standards.
>
> Different types of honey fraud can be achieved through:
>
> dilution with different syrups produced, e.g. from corn, cane sugar, beet sugar, rice, wheat, etc.;
>
> harvesting of immature honey, which is further actively dehydrated by the use of technical equipment, including but not limited to vacuum dryers;
>
> using ion-exchange resins to remove residues and lighten honey colour;
>
> masking and/or mislabelling the geographical and/or botanical origin of honey;
>
> artificial feeding of bees during a nectar flow.
>
> The product which results from any of the above-described fraudulent methods shall not be called 'honey' neither the blends containing it, as the standard only allows blends of pure honeys.
>
> Apimondia have rejected the idea of methods being developed with the intention of artificially speeding up the natural process of honey production through an unnecessary intervention of humans and technology that may lead to violations of honey standards. Honey fraud defaces honey's image of being a natural product. It also affects the consumer as they are not getting the product they are paying for as a result threatening food safety and security.

It's worth noting that actively dehydrating honey constitutes fraud. This is a practice that I've heard is undertaken by some beekeepers to upgrade what would be Baker's Honey by reducing the moisture content.

The science of honey testing for fraud is now very sophisticated indeed, including pollen analysis, DNA meta-barcoding, Nuclear Magnetic Resonance, and liquid chromatography coupled with high resolution mass spectrometry (LC-HRMS) and carbon isotope ratio analysis. However, the fraudsters are apparently producing sophisticated syrups profiled to match specific honeys in an attempt to evade detection, and, as stated, counterfeiting of honey is resulting in serious economic, environmental, and potentially food security issues. The National Honey Monitoring Scheme are building a databank of multiple gene analysis of different honey samples, together with a profile of the carbohydrate content, to enable a certification process.

Nuclear magnetic resonance spectrometer © Sarayut Watchasit | Dreamstime.com

There are other crimes possible in connection with food, and some of these are:

**Adulteration**: Reducing the quality of a food product through the inclusion of a foreign substance, with the intention either to make production costs lower, or apparent quality higher.

**Substitution**: Replacing a food product or ingredient with another substance of a similar but inferior kind.

**Misrepresentation of origin, quality, provenance or benefits:** The marketing or labelling of a product so as to inaccurately portray its quality, safety, benefit, origin or freshness.

**Document Fraud**: The use of false or misappropriated documents to sell, market or otherwise vouch for a fraudulent or substandard product.

## Sampling programmes

Any food may be sampled by, for instance, a Local Authority and County Councils, as part of their food-sampling programme. You'll also see references here to samples taken by the FSA, or the VMD, where the results have been published online and so are readily accessible.

Samples by Local Authorities can be either informal, or formal.

### *Informal samples*

These simply involve an officer purchasing a product in a shop – they don't declare their identity, or that the food is being sent for analysis. The sample would be bagged up, and then sent off to the Lab. for analysis. The results would probably be shared with the shop or the producer, although if a problem were found it may result in further investigation or action.

### *Formal samples*

These are more complex, and prosecution could result if the results showed contraventions. Essentially the Officer takes a sufficiently large sample of the subject, and divides it into three separate parts. These are then sealed, and one taken for analysis; the second is given to the retailer, so they can have it analysed if necessary, and the third is retained sealed, as a reference sample.

If the analysis of the first sample is satisfactory, then this will be communicated back to the supplier, and no further action taken; if the analysis is unsatisfactory, the supplier is told and they can have their sample analysed, at their own cost, to verify the first analysis. Should the results of the two analyses be inconsistent, then the third, reference, sample is sent to the Government Chemist for repeat analysis, and that analysis is regarded as authoritative.

## Analysis

There is a huge range of tests that can be conducted on honey, to detect contamination and fraud – one laboratory reports that their most requested analyses of honey include:

| | Parameters according to EC Honey Directive | | | |
|---|---|---|---|
| sugar content | water content | Diastase-activity (Schade, Phadebas and nitrophenol method) | Hydroxymethylfurfural (HMF) |
| water-insoluble substances | electrical conductivity | GMO (genetically modified organisms) | acidity |
| **Bee medicines** | | | |
| Amitraz | Chlorfenvinphos | tau-Fluvalinat | Flumethrin |
| Coumaphos | Brompropylate | 4,4'-Dibrombenzophenon | |
| **Antibiotics** | | | |
| Tetracycline | Streptomycine | Chloramphenicol | Fluorchinolones |
| Sulfonamides | Tylosine | Makrolides | Nitroimidazoles |
| Nitrofuran Metabolites | | | |
| **Pesticides** | | | |
| Neonicotinoids | Glyphosate | | |
| **Tests for adulteration:** | | | |
| NMR | 13C Isotope analysis (EA-IRMS + LC-IRMS) | Detection of adulteration marker psicose | Determination of caramel colour |
| honeyforeign amylase | honeyforeign invertase | Detection of beet sugar syrup marker (SM-B) | Oligosaccharides from starch based syrups |
| thermoresistant Amylase | Foreign a-amylases | Detection of rice syrup marker substance (SM-R and TM-R) | |
| **Microbiology** | | | |
| Aerobic, mesophilic bacteria | Salmonellae | Staphylococcus | Enterobacteriaceae |
| Yeasts | Moulds/spores | mesophile sulfite reducing Clostridia | Listeria |
| E. Coli | Coliform Bacteria | | |
| **Heavy metals** | | | |
| LeadIron | Arsenic | Cadmium | Mercury |
| **Others** | | | |
| Pollen analysis to determine origin Pyrrolizidin alkaloids | | | |

Adverse results are likely to be investigated, either by the organisation taking the sample, or being referred on – in the case of samples containing lead and PDB, taken by the VMD, the beekeepers in question were visited in an attempt to identify where and how the contamination could have arisen.

## Enforcement

Often enforcement is seen as being "taken to Court", but there are a range of other enforcement routes that can be taken. Enforcement needs to be "proportionate" to the nature and scale of the problem, so a trivial matter is likely to be treated informally, where something more consequential (or a failure to change after informal action) will be formal. Serious matters could always result in formal action (service of a Notice or Prosecution).

The Food Law Code of Practice describes a Hierarchy of Enforcement, and the options include educating food business operators, giving advice, informal action, sampling, detaining, and seizing food, serving Hygiene Improvement Notices/Improvement Notices, Remedial Action Notices, Hygiene Prohibition Procedures/Prohibition Procedures, and Prosecution.

Education includes making available training as described above, and giving advice will often be by simply discussing the production process during the inspection, any shortcomings being identified, and making suggestions or comments as they go along.

Often an informal letter will be sent summarising the inspection and listing the suggested actions, with a timescale for them to be completed, with a follow-up visit programmed for review.

Sampling food may be done during the inspection; this would simply be taking something, often by agreement, so that it can be subject to more detailed investigation or checks, and it should be seen as being separate from the sampling discussed in the appropriate chapter.

Detaining and Seizing food would both be in the light of concerns about the safety of that specific food – and are likely to be followed by destruction of the food, either by agreement with the owner or following an order of the Court. This would occur if the food was found to be non-compliant or unsafe, and no compensation would be payable. In one unpublished case a few years ago, a year's entire honey crop – 200 hives' worth – was in danger of being destroyed, as PDB had been detected when a sample had been analysed.

Hygiene Improvement Notices and Remedial Action Notices would require specific steps to be taken within a set period of time, and failure to comply with the Notice is the Offence.

Prohibition procedures would involve prohibiting a particular process, use of equipment or premises. If not informally agreed, then formal action would be taken including an application to the Courts.

Premises can be closed, prohibiting their use until essential steps have been taken.

Finally, a prosecution may be undertaken. This is an option against any absolute offence, but there are a couple of tests that should be passed before a decision to prosecute is taken. The first is the "Evidential Test" – simply is there sufficient evidence for there to be a realistic prospect of conviction – and the second is the "Public Interest Test": there are several factors to be considered here, including the seriousness of the offence, the level of culpability, circumstances and harm caused.

When appropriate, non-food safety charges can be brought – in the case of the Bakers (above) the charge was "obtaining property by deception" – a fraud charge.

> In 2017 Mohammed Kuddus and Harun Rashid[42] were sentenced to prison for manslaughter following the death of Megan Lee from an allergic reaction to nuts contained in a take-away that she had eaten. In 2016, Mohammed Zaman[43] was sentenced to six years in prison for manslaughter for a similar incident. John Croucher[44] was sentenced to four months, suspended for 12 months, and fines totalling £12,000 were imposed after an Elizabeth Neuman died from food poisoning – Croucher had undercooked a shepherd's pie and not checked the temperature with a probe thermometer before serving the meal. Murray McGregor[45] was convicted under the Veterinary Medicines Legislation for importing antibiotics and dosing his bees with them.

The Food Law Code of Practice also requires that a graduated approach be taken, with factors such as the risk to health, and whether there have been fraudulent or deceptive practices being considered. Additionally, there is a "Code of Practice for Crown Prosecutors" which also gives guidance, and separately sentencing guidelines in the event of a Conviction.

It should be noted that an offence is committed by failing to comply with a Notice – should there be a prosecution, the Court will not look at the requirements of the Notice, just that it was served (and not necessarily that it was received), and not complied with within the set time.

**Rights and powers of Officers**

Authorised officers of the Local Authority will have a photo-identity card and a Certificate of Authority with them whilst working, and this can be requested.

In case of a domestic premises, then 24 hours' notice is required.

Their right of access extends to all parts of the premises used by the business.

The inspectors can inspect premises, processes and documents, they can seize articles or equipment, documents, and take samples of food for analysis, take photos and videos and other evidence as they see fit.

> Samples have been known to include, for instance scrapings of grease from an unwashed floor, with mouse droppings in the compacted grease.
>
> When presented in Court, this was compelling evidence.

Failure to provide access on request is an offence, and the officer is entitled to police assistance where necessary.

There is a duty on the food business operator to comply and not to obstruct the Officer.

**The inspection process**

The inspection is an information-gathering process, and in addition to reviewing the premises and the equipment, expect to be asked to see documentation relating to food hygiene practices, and perhaps to go through it with the Officer, explaining and clarifying. Documentation, as in the SFBB packs, completed, would be seen as very good indeed.

Safer Food Better Business - a good management system for small and medium enterprises.

Hive records (inspections, splits etc.) do not really form part of the documentation but could be made available – though they may well contain relevant information, such as Veterinary Medicine treatments, feeding, and yields.

> There's no prescribed form for records: however in the case of Veterinary Medicines, there is a card which can be downloaded from the Medicines page of BeeBase,[46] and – if this is completed and retained for five years, compliance is assured.

Records of cleaning (check lists), receipts for items and equipment, HACCP records, Batch records for honey and other products could all be relevant, as could invoices and receipts. Records of bulk / wholesale purchase or sale – if either have been done – should also be available.

It's unlikely that the Officer will go through the records with a fine-tooth comb, but may ask for clarification of some information, or ask if particular records are kept, and if not, then why not. The only statutory requirement relates to Veterinary Medicines, where the requirement is that they are kept for five years. Higher scores in the Food Safety Rating Scheme will be awarded for good documentation.

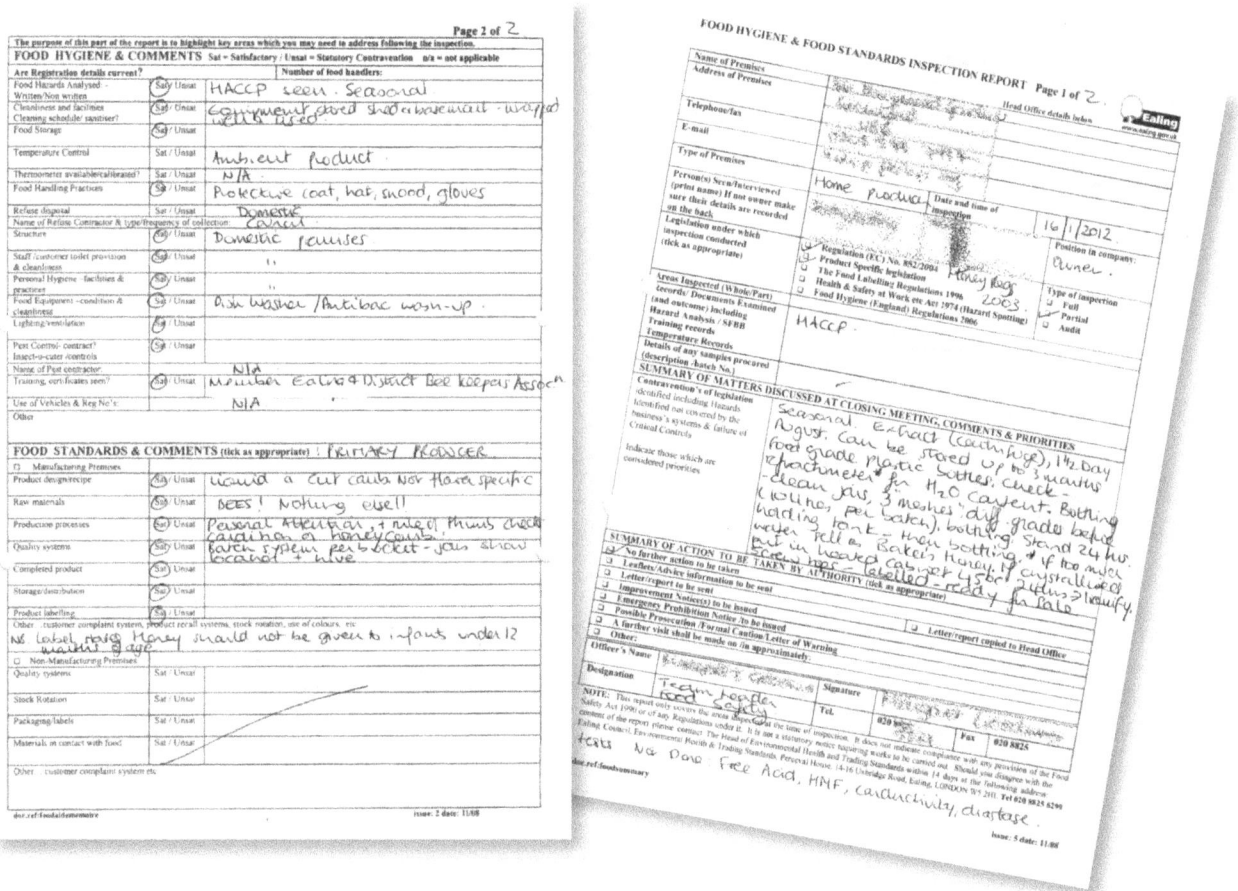

## GM crops

Under EU legislation, GM crops are banned. Although the UK has left the EU, the legislation remains in force – see the section on Brexit above.

British Honey Importers and Packers Association (BHIPA) has a statement on GM crops in Honey[47] which states:

> "The European Court of Justice (ECJ) issued a judgement in September 2011 that honey containing GM pollen falls within the scope of EU Regulation 1829/2003 (GM Regulation). Consequently the pollen in such honey must have a specific GM food authorisation before the honey can be placed on the market."

The statement asserts that to date their testing has not found any GM pollen contamination, and that they will continue to sample for this.

As GM crops are banned in the UK, pollen from GM Crops should not be present in UK Honey.

EU law requires that if GM pollen were present at levels of 0.9% by volume in honey, then it would need to be declared on the label; however as pollen never exceeds 0.5% of volume, in effect there will never need to be such a declaration.[48]

## Due Diligence

There is a defence against an absolute offence, which is "due diligence". The actual phrasing is:

> "prove that he took all reasonable precautions and exercised all due diligence to avoid commissioning of the offence by himself or a person under his control."

Due diligence is a phrase that's applied in a wide variety of circumstances – when you buy property, for instance, your Solicitor has to exercise due diligence to ensure that the Title is the vendor's to sell and the transaction goes through properly, your financial advisor's that they are giving accurate advice, and so on.

It is something that is hard to define – Wikipedia gives:

> "due diligence is the only available defence to a crime that is one of strict liability ... Once the criminal offence is proven, the defendant must prove on balance that they did everything possible to prevent the act from happening. It is not enough that they took the normal standard of care in their industry – they must show that they took every reasonable precaution."

> "In the United Kingdom, 'proper use of a due diligence system' may be used as a defence against a charge of breach of regulation ..."

Words such as *due, reasonable, normal,* and *necessary* are subjective in their interpretation. So a "due diligence" defence will need to be made, and argued in front of a Court, with supporting evidence – and what evidence can you produce? This is where documentation – records – are important.

Take, for instance, HMF – should a sample be taken and an excessive level of HMF be found, then due diligence could be evidenced by records of where and for how long the honey was stored before sale (in a warm cupboard would not be ideal!), and whether, or how it was heated, what temperature was reached, and how long it was kept at that temperature for. There's a specimen Batch Record Sheet in Appendix 3.

The HMF limit of 40mg/kg applies at the time of sale, so storage instructions (keep in a cool place) given to the retailer would help the beekeeper's case, and the durability period (best before, say, six months) would limit the time, so limiting the development of HMF.

Another example could be the purchase of honey in bulk from a neighbouring beekeeper, and then you pack it for retail sale. The product is sampled and found to be adulterated with sugar syrup. You, and your customers have been had. The accusing finger points at you – it is your name on the jar. But, using batch numbers and your immaculate set of records, you'd be able to identify that the fraudulent batch of honey had been acquired in bulk, where and when you got it from. Your records, here could include invoices, receipts, maybe even other documents such as your bank statements showing a BACS or cheque payment for the honey, emails, texts or other messages setting up the transaction could all help demonstrate due diligence.

## Organic

There is very strict control on the use of the word "organic" in connection with any product – the consumer is paying a premium price for a product that they fully expect to be produced to particular, and high standards.

Over the years I've seen honey labelled, or presented, as being organic in some way – and there are often online discussions about the use of the word.

The use of the word *organic* is protected and can only be used in certain, and very tightly defined, circumstances.

The first requirement is that you need to be registered with a UK Approved Control Body, and work to their standards. A fee is payable, and there will be checks.

There is a list of UK Approved Control Bodies in the Government Website.[49]

One of the best-known Control Bodies in the UK is the Soil Association, and their requirements for registration can be found on their website. Specific requirements include:

- a conversion period of 12 months
- local ecotypes of Apis mellifera are kept (no imported queens)
- Stocks are increased either by splits from your own or by bringing bees in from other organic apiaries.
- not keeping organic and non-organic bees together
- managing the bees to prevent disease
- notification of the Control Body of treatments; additional provisions apply if the treatment involves any chemically synthesised allopathic veterinary treatments
- not to mutilate bees, e.g. by clipping the queen's wings
- leaving sufficient stores for the winter; feeding only if essential, use organic sugar or syrup; keep detailed records of this
- Emergency feeding needs to be authorised in advance by the Control Body, and again with organic sugar or syrup.
- Some regions may just not be suitable for organic honey production: the Control Body will have listed these and hives must not be established in these areas.
- Apiaries must be in areas of organic food production, or unmanaged wild areas, or areas where crops are managed with "low environmental impact methods and which cannot significantly affect the organic description of beekeeping," and far enough from sources that could lead to contamination of the products. A map of the area around each apiary needs to be supplied to the Control Body.
- Hives need to be made from natural materials giving no risk of contamination of the environment, the bees or the products.
- Cleaning hives requires specified materials or methods.
- Wax (e.g. foundation) has to be from organic sources, and non-organic wax / foundation to be removed as part of the hive's conversion to organic production.
- Harvest must not involve chemical repellents, and must not involve destroying bees or brood.

Misusing the word organic in marketing is taken seriously by the Courts – in 2005, Stephen Sains, a butcher in Richmond, Surrey was fined £6,000 for falsely labelling food;[50] and Andrew Portch, a Somerset Farmer, was fined £3,000 for offences including using organic certification without the required accreditation. Company Director Andrew Stansfield, of Northamptonshire, was sentenced to 27 months. His partners in crime, Kate Stansfield and Russell Hudson, both received suspended sentences and 150 hours of community work. All three pleaded guilty to selling conventionally grown food as organic. The business is believed to have misleadingly described over £500,000-worth of food over a period of five years.[51]

## Quantities and Weighing

A standard, 1lb honey jar.

Historically, honey had to be sold in prescribed weights under the Weights & Measures Act 1985. These were 57g (2oz), 113g (4oz), 227g (8oz), 340g (12oz), 454g (1lb) or multiples of 454g (1lb). However, that provision has now been repealed[52] and pre-packaged honey may now be sold in any weight.

Special jars are made to accommodate 454g (1lb) or 227g (1/2lb) of honey; these are still in production, they are the standard for most classes at honey shows, and are frequently seen on sale.

They are marked with a "fill line" to give the appropriate weight of honey, although it is not sufficient to rely on this to ensure your customers are being supplied with the required weight of honey: the filled jars must be weighed on Government Stamped scales.

As the prescribed weights are no longer obligatory, any jar can be used – suppliers produce a huge range of jars, round, square, hexagonal, octagonal, and novelty shapes which may be used, in both plastic and glass. Most of these will be standard food jars used generically for many products, and as the density of honey is greater than that of water, they do not hold a "standard" weight of honey and look full. For instance, what would be seen as a 12oz jar is sold as a 340ml (millilitre) jar, and needs about 360 grammes of honey to fill it so its appearance is satisfactory to the customer. 340 grammes equates to 12 ounces, so using Imperial weights, you would have labelled that 340g 12oz, where you could label it as 360g so the customer knows the exact weight they are getting.

Cut Comb and Sections are not sold by weight, but simply on their appearance.[53] You may wish to mark the weight on the packaging for the customer's information, but that is not a legal requirement. I've not been able to identify a specific exclusion for Sections, which are of course of unpredictable and uncontrollable weight, by their very nature – however, the Honey Regulations (England) Regulations use the description:

> "comb honey" means honey stored by bees in the cells of freshly built broodless combs or thin comb foundation sheets made solely of beeswax and sold in sealed whole combs or sections of such combs;

so Sections, it seems, are included within the definition of Comb honey.

**Weighing Systems**

There are two systems for weighing – one is the "Minimum System" where every container contains at least the declared weight; the other is the "Average System" under which the average content for all packs will be the declared weight, and while some will be overweight, others may be slightly underweight. There are strict controls here – known as the "Three Packers Rules" – which require random sampling to ensure that the system is complied with, only a small number are below the "Tolerable Negative Error" (TNE) (which is defined in law) and that no package is short by more than twice the TNE.[54] If using the Average system, optionally the weight can be accompanied by the "℮" symbol.

**Weighing equipment**

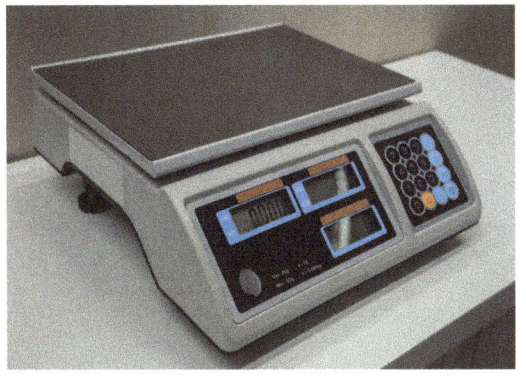

"Government stamped" scales **must** be used when weighing your produce, and keep records of the scales that you use. The scales need to be verified periodically for accuracy and conformity, the frequency varying from 6 months to 5 years, depending on the use that the scales are put to – if they are used a lot, or subject to rough handling, then they'd need to be inspected more frequently. Keep a record (certificate or invoice) of the date of the check – typically there would be some kind of certificate or report supplied.

Jars from the same batch do not vary much in weight in my experience, so I take an average weight for six jars and lids and then use that for my Tare weight (the difference between the Gross weight and the Nett Weight) when doing my check weighing.

## Durability

This is the shelf life of a food, and marking it is required on all foodstuffs.

Honey is perceived as a very "safe" food and, apocryphally, honey found in the Pyramids, thousands of years old, was "perfectly edible". Beekeeping was certainly practised in Ancient Egypt, but it is a mistake to use this as a basis for a claim that it does not spoil and can be kept forever. In fact, it seems that, in 1904, the contents of a jar were mistakenly described as honey, and although later corrected the myth continues.[55]

There is other evidence that honey has been harvested for 3,500 years, but the honeypots that revealed this contained only traces.

There are two formats for durability markings:

**Use by:** this is for perishable foods, and gives a date after which the food may be unsafe. The durability relies on certain conditions – a highly perishable food will keep longer in a refrigerator, or much longer deep-frozen. If so, the label must specify the storage conditions and the use-by date will be set based on correct storage.

You may want to use this on compound food – honey cakes or bread, honey mustard and such – which will deteriorate after just a few days.

**Best Before:** for less perishable foods, where the product will deteriorate usually over a longer period of time, not actually become unsafe at the end of the period.

> It is surprising what can affect durability – in 1989, there was an outbreak of botulism, which was traced back to a hazelnut yoghurt.
>
> The yoghurt had been in production[56] for some time without any problems, but then the hazelnut puree that it contained was reformulated, replacing sugar with an artificial sweetener, aspartamine. This changed the Water Activity and coupled with inadequate heat treatment, the bacterium Clostridium botulinum was able to proliferate, producing its toxin.
>
> 27 patients were affected.

Setting the Use By and Best Before dates requires an awareness of the nature of the food, how it will deteriorate, and what could happen if consumed beyond its durability. The factors that affect it include storage temperature, water availability, pH (acidity) and the presence (or absence) of oxygen. Packaging can have a big influence – ripe honey hermetically sealed in a screw-top jar will not ferment: the same honey sold unsealed will absorb atmospheric moisture and ferment.

In the case of a cooked meat product, bacterial spoilage may well take place and the bacteria produce toxins or cause food poisoning or a food-borne illness, so a strict "use by" date will be set, with a requirement that the product be kept under refrigeration until it is to be consumed – the "cold chain".

In the case of a honey biscuit, the shelf life may be quite long and the spoilage may be that the fats become rancid, making the product unpalatable. The product may dry out if it is stored inappropriately (so the label may include "store in an airtight container"). So a "best before" date of several months may be acceptable.

If you were to be asked, "How do you determine the durability?" of any product, you need to be able to answer, with a sound basis for the answer. Commercially, laboratories may be asked to undertake tests on food, to assess their durability, and give a recommendation accordingly. This may be beyond the resources of a small-scale producer, so comparisons with similar commercial products may be helpful – but bear in mind that factory-produced goods may have preservatives, antioxidants and the like included to extend the shelf-life. Also, that commercial packaging materials or techniques may have properties that preserve the product, which are not available to a smaller-scale producer.

Considering honey specifically, with your refractometer you can confirm it is ripe, so should not ferment; it is packed in a hermetically-sealed jar; what could its shelf life be? In its guidance leaflet "Selling Honey, Complying with the Law", the BBKA suggested a "Best Before" date of two years. The same was suggested in a recent government webinar.[57]

Consider though that, although you can be sure that the honey will not ferment, you do not know what is happening to the Diastase, Invertase, and HMF, and the regulations apply at the time of sale to the final consumer. Perhaps a reasonable approach is to assume that the honey is stored at ambient temperatures – 20°C is a reasonable average – so refer to the various graphs available to show how these change over time and with temperatures, allow a bit of a safety margin, and use that to set your "Best Before" date.

Research[58] suggests that based on Diastase activity, 15 to 20 months is reasonable.

## Caveat

Only the courts can interpret the law – this is to be seen as guidance so you can decide how the legislation applies to you and your business model.

This is a summary only – hopefully the main points.

## Appendix 1

# Hazard Analysis and Critical Control Point for Liquid Honey

{Business or Producer Name} Date: {dd/mm/yy}  Date of next Review {dd/mm/yy}  Page 1/4

| Hazard | Step / Stage/ Location / source | Type | Nature / source of hazard | Critical Limits | Controls | Monitoring / testing if necessary | Corrective Action |
|---|---|---|---|---|---|---|---|
| Contamination | Hive | Chemical | Preservatives (various) | Zero | Ensure woodwork of hive treated with non-toxic, odour-free treatments if possible approved for use with hives by their suppliers. | | Withdraw from use if so treated. |
| | | Chemical | Veterinary medicines | Zero | Use only approved medications Fully comply with treatment regime Mark supers of relevant hives so they are identifiable & keep separately Extract honey from treated hives separately and wash equipment afterwards to prevent contamination. Label buckets containing extracted honey from treated hives to prevent accidental sale and keep separately Set aside honey from treated hives for the duration of the withdrawal period Allocate a different lot number to honey from treated hives. | Keep records as required by legislation | Withdraw honey from sale / observe Withdrawl periods where relevant |
| | | Chemical | Toxins from forage (pesticides, toxic nectar, environmental toxins) | | Currently no controls needed in the UK. | Monitor bee press etc. for information and news | Review HACCP analysis in light of new information as and when received. |
| | | Physical | Contamination by wax moth webbing and faeces | Zero | Minimise wax moth. Process quickly / minimise storage times Freeze comb to be used for cut comb, chunk honey & sections to kill eggs and larvae. Filter / strain honey to remove contaminants. | Visual inspection | Withdraw damaged comb from use. |
| Contamination | Hive, transport, storage | Physical / Bacterial | Physical contamination, dirt etc | Zero | Ensure honey supers are not placed on the ground when manipulating bees. Ensure barrows, vehicles etc are clean when transporting supers. Take care when transporting supers. Line vehicles with clean sheeting or trays. Store unused equipment in hygienic conditions. | Visual inspection. | Withdraw soiled combs from extraction process. |

# Hazard Analysis and Critical Control Point for Liquid Honey

{Business or Producer Name}  Date: {dd/mm/yy}  Date of next Review {dd/mm/yy}  Page 2/4

| Hazard | Step / Stage / Location / source | Type | Nature / source of hazard | Critical Limits | Controls | Monitoring / testing if necessary | Corrective Action |
|---|---|---|---|---|---|---|---|
| | | Chemical | Contamination by PDB / Naphthalene | Zero | Use only approved materials and techniques for wax moth control / storage of comb. Withdraw from use any honeycomb that could be contaminated with either chemical from previous treatments. | Laboratory testing if contamination is possible. | withdraw contaminated honey from sale. |
| Contamination | Hive | Chemical | "Hive cleaners" and non-medicinal curative substances, tonics etc | Zero | Only use proprietary / approved products. Follow manufacturer's instructions. Record use as if Veterinary Medications, and comply with withdrawal periods | Laboratory Testing may be necessary if manufacturers have not undertaken testing and published results showing that there is effect on honey. | Withdraw honey from sale. |
| | Processing and preparation | Bacterial / microbiological | Food handlers | | Trained or instructed and supervised Comply with relevant legislation and good practice etc. | Basic food hygiene course passed, Supervision | |
| | | Bacterial/ Physical/ Chemical | Food premises | | Comply with food safety legislation requirements re cleanliness, facilities, space, construction, sanitary accommodation, washing facilities, pest free, etc | Check labelled as food-safe, retain records as to origin. | |
| Contamination | | Chemical / bacterial / physical | Equipment | | Stainless steel, food grade plastic etc. to be used, **galvanised, tin plate, or lead-soldered equipment not to be used under any circumstances.** Purpose-made filtration equipment only to be used. Equipment to be cleaned with food-safe detergent / sterilant prioor to use. | Check labelled as food-safe, retain records as to origin. | Replace any defective |
| | | | | | Ensure clean and in good condition | Visual inspection | Clean and rinse with eg Milton solution, allow to dry before use. Use disposable paper towels for drying |

# Hazard Analysis and Critical Control Point for Liquid Honey

{Business or Producer Name}  Date: {dd/mm/yy}  Date of next Review {dd/mm/yy}  Page 3/4

| Hazard | Step / Stage / Location / source | Type | Nature / source of hazard | Critical Limits | Controls | Monitoring / testing if necessary | Corrective Action |
|---|---|---|---|---|---|---|---|
| Contamination | | Chemical | Contamination from cleaning chemicals | None | Food-safe detergents etc., only to be used. Rise equipment with clean water; dry thoroughly. Store cleaning materials securely away from working area | Visual inspection | |
| | | Chemical / allergen | From latex gloves | None | Only use vinyl or other food-safe gloves when extracting handling and packing honey | None | Ensure latex gloves not used. |
| | | Avoid use of metal or plastic scourers | Removal of pollen | Statutory "excessive" | Avoid using very fine mesh to strain honey prior to packing. NB mesh size greater than 100 microns would be acceptable | None | |
| | Bottling | Physical | Glass chips from damage to jars | none | Handle jars carefully to prevent cracking and chipping. Reject damaged jars/lids. Avoid overtightening jar lids to avoid damage (hand tighten only) | Visual inspection before filling and at time of sale. | Withdraw damaged jars. |
| | | Physical / Bacterial | Contamination from jars | | Store unfilled jars and lids in hygienic conditions. Wash, rinse and dry jars and lids thoroughly (dishwasher best) before filling | Visual inspection | |
| | | Chemical | Plastic containers (jars / cut comb trays etc.). | | Ensure food grades | Check for food-safe labelling; certification. retain record of where purchased from. | |
| | | Various | Accidental damage or malicious contamination | nil | Handle containers with appropriate care. Use of anti-tamper labels to make tampering apparent. | | |

# Hazard Analysis and Critical Control Point for Liquid Honey

{Business or Producer Name}   Date: {dd/mm/yy}   Date of next Review {dd/mm/yy}   Page 4/4

| Hazard | Step / Stage/ Location / source | Type | Nature / source of hazard | Critical Limits | Controls | Monitoring / testing if necessary | Corrective Action |
|---|---|---|---|---|---|---|---|
| Spoilage | Excessive moisture in the honey | Microbiological | Fermentation | As Honey Regs, - 20% for Honey, up to 25% for Bakers Honey. | Ensure all equipment thoroughly dry before use ensure food handlers dry their hands effectively after washing. Only extract from fully ripe combs. Work in as dry (low relative humidity) an environment as possible. Store extracted honey in hermetically sealed containers. Tighten jar lids fully when packaging for sale. Fit anti-tamper seals. Label to include storage instructions. | Refractometer or other test. Visual inspection. "shake test" | Sell honey with moisture content greater than 20% as "Bakers Honey". |
|  | Processing and storage | Chemical | Excessive HMF | Statutory 40 mg/kg | Avoid heating honey unnecessarily. Minimise heating when it is necessary – use thermostat and timeswitch to give Maximum temperature of 42C and duration 24 hours maximum | Record maximum temperature and duration of heating Laboratory Testing | Sell honey likely to have excess HMF as "Bakers Honey" / withdraw from supply and use personally or as an ingredient. |
|  |  | Chemical | Loss of Enzyme activity | Statutory ("Excessive") but no numerical criteria | Store in a cool place. Minimise storage – stock rotation. Use best before dates. |  |  |
|  |  | Physical / chemical / bacterial | General deterioration |  | Stock rotation. "Best before" date. | Visual inspection at time of sale. |  |
|  |  | Loss of enzymes | Micowave radiation destroys enzymes | "significant reduction" | Avoid use of microwave oven to liquefy crystalised honey | None | Label and sell as Baker's Honey if microwave energy used to liquefy. |
| Microbiological | Consumption | Microbiological | Presence of C. botulism spores |  | Normal good hygiene practice. Include label on jar "honey should not be given to infants under 12 months of age" or similar | Visual inspection | Withdraw from sale until appropriately labelled |
| Allergens | Consumption | Chemical | Presence of allergens in product | Zero unless identified on label | None of the listed allergens are present in honey. | None | Nil |

# Appendix 2

**SCHEDULE 1 of The Honey (England) Regulations 2015**
**Compositional criteria**

1. The honey consists essentially of different sugars, predominantly fructose and glucose, as well as other substances such as organic acids, enzymes and solid particles derived from honey collection.
2. The colour varies from nearly colourless to dark brown.
3. The consistency can be fluid, viscous or partly or entirely crystallised.
4. The flavour and aroma vary but are derived from the plant origin.
5. No food ingredient has been added, including any food additive.
6. No other additions have been made to the honey except for other honey.
7. It must, as far as possible, be free from organic or inorganic matters foreign to its composition.
8. It must not—
    (a) have any foreign tastes or odours;
    (b) have begun to ferment;
    (c) have an artificially changed acidity;
    (d) have been heated in such a way that the natural enzymes have been either destroyed or significantly inactivated.
9. Paragraph 8 does not apply to baker's honey.
10. No pollen or constituent particular to honey may be removed except where this is unavoidable in the removal of foreign inorganic or organic matter.
11. Paragraph 10 does not apply to filtered honey.
12. The additional compositional criteria set out in the following table apply—

| Criteria | Amount |
|---|---|
| **Sugar content** | |
| 1.—(1) Fructose and glucose content (sum of both)— | |
| (a) blossom honey | not less than 60g/100g |
| (b) honeydew honey and a blend of honeydew honey with blossom honey | not less than 45g/100g |
| (2) Sucrose content— | |
| (a) all honey except for honey specified in paragraph (b) or (c) | not more than 5g/100g |
| (b) false acacia (*Robinia pseudoacacia*) honey, alfalfa (*Medicago sativa*) honey, Menzies Banksia (*Banksia menziesii*) honey, French honeysuckle (Hedysarum) honey, red gum (Eucalyptus camadulensis) honey, leatherwood (*Eucryphia lucida, Eucryphia milliganii*) honey, Citrus spp. honey | not more than 10g/100g |
| (c) lavender (Lavandula spp.) honey, borage (*Borago officinalis*) honey | not more than 15g/100g |
| **Moisture content** | |
| 2. Moisture content— | |
| (a) all honey except for honey specified in paragraph (b), (c) or (d) | not more than 20% |
| (b) honey from heather (Calluna) | not more than 23% |
| (c) baker's honey except for baker's honey from heather (Calluna) | not more than 23% |
| (d) baker's honey from heather (Calluna) | not more than 25% |
| **Water-insoluble content** | |
| 3. Water-insoluble content— | |
| (a) all honey except pressed honey | not more than 0.1g/100g |
| (b) pressed honey | not more than 0.5g/100g |
| **Electrical conductivity** | |
| 4. Electrical conductivity— | |
| (a) all honey except for bell heather (Erica) honey, chestnut honey, eucalyptus honey, honeydew honey, lime (Tilia spp.) honey, ling heather (*Calluna vulgaris*) honey, manuka or jelly bush (Leptospermum) honey, strawberry tree (*Arbutus unedo*) honey and tea tree (Melaleuca spp.) honey | not more than 0.8mS/cm |

| | | |
|---|---|---|
| (b) blends of honeys to which paragraph (a) applies | not more than 0.8mS/cm | |
| (c) honeydew honey | not less than 0.8mS/cm | |
| (d) blends of honeydew honey except blends of that honey with bell heather (Erica) honey, eucalyptus honey, lime (Tilia spp.) honey, ling heather (*Calluna vulgaris*) honey, manuka or jelly bush (*Leptospermum*) honey, strawberry tree (*Arbutus unedo*) honey and tea tree (Melaleuca spp.) honey | not less than 0.8mS/cm | |
| (e) chestnut honey | not less than 0.8mS/cm | |
| (f) blends of chestnut honey except blends of that honey with bell heather (Erica) honey, eucalyptus honey, lime (Tilia spp.) honey, ling heather (*Calluna vulgaris*) honey, manuka or jelly bush (*Leptospermum*) honey, strawberry tree (*Arbutus unedo*) honey and tea tree (Melaleuca spp.) honey | not less than 0.8mS/cm | |
| **Free acid** | | |
| 5. Free acid— | | |
| (a) all honey except for baker's honey | not more than 50 milli-equivalents acid/kg | |
| (b) baker's honey | not more than 80 milli-equivalents acid/kg | |
| **Diastase activity and hydroxymethylfurfural content** | | |
| 6. Diastase activity and hydroxymethylfurfural content (HMF) determined after processing and blending— | | |
| (i) all honey except baker's honey and honey specified in sub-paragraph (ii) | not less than 8 | |
| (ii) honey with a low natural enzyme content (e.g. citrus honey) and an HMF content of not more than 15mg/kg | not less than 3 | |
| (b) HMF— | | |
| (i) all honey except baker's honey and honey specified in sub-paragraph (ii) | not more than 40mg/kg | |
| (ii) honey of a declared origin from a region with a tropical climate and blends of these honeys | not more than 80mg/kg | |

## Appendix 3

# Honey batch record

| | | | | | | | |
|---|---|---|---|---|---|---|---|
| Hive # | | Super # | | Date on hive | | Date off hive | |
| Apiary Location | | | | Likely Crop | | | |
| Approx weight (gross) | | | | Batch # | | | |
| Refractometer | Make/model | | % Sugar | | % H$_2$0 | Ambient Temp | °C |
| Selected packaging | | 454 g | 340 g | 226g | Other (SAY) | ◯  ⬡  ▢ | Other (SAY) |
| Packing material | | container | Glass | Plastic | Other (SAY) | Lid  Metal  Plastic | Other (say) |
| Packing purchased from | | | | | | Name, address, lot number, keep the receipt | |
| No. of jars filled | | | | Date extracted | ddmmyy | Date bottled | ddmmyy |
| Heat treatment | | Duration | | hh:mm | | Max Temp | °C |
| Other processing (eg creaming etc.) | | | | | | Continue with detail overleaf if necessary | |
| Scales Verification | | Calibration check | Expected | actual | Scales used | Make/model | Date of check  ddmmyy |

### Packers Rules

| | | | |
|---|---|---|---|
| Weight, 6 jars plus lids | | Weight, individual jar | |
| Weight of fill. | | Minimum weight to be acceptable | |

### Check Weights

| | | | | | |
|---|---|---|---|---|---|
| | | | | | |
| | | | | | |
| | | | | | |
| | | | | | |
| | | | | | |
| | | | | | |
| | | | | | |

If a compound food, list ingredients here & continue overleaf as necessary

| Ingredient (e.g. Egg, flour, milk etc) | source (e.g. shop name) | batch no | allergen y/n | receipt? |
|---|---|---|---|---|
| | | | | |
| | | | | |
| | | | | |
| | | | | |

| | | | |
|---|---|---|---|
| Date | / /20 | Signed | |
| Name | | | |
| Address | | | |
| Trading Name | | | |

## Credits

All images copyright Andy Pedley except where stated.

Thanks to Cherwell District Council for permission to use the food room images, Martyn Butt in the front cover image, Ann Fox, the model Food Handler and Frances Burnett, for the cartoon on page 76.

# Endnotes

1. https://www.breitbart.com/news/uk-launches-post-brexit-review-of-eu-laws/
2. http://redwall.unchecked.uk/wp-content/uploads/sites/2/2020/12/A-view-from-the-Red-Wall.pdf
   https://demos.co.uk/wp-content/uploads/2021/03/Food-in-a-Pandemic.pdf
3. National statutory surveillance scheme for veterinary residues in animals and animal products: 2020, 2020
4. https://www.food.gov.uk/business-guidance/businesses-that-supply-or-produce-food-on-the-move
5. https://www.tentamus.com/news/hmf-honey-qsi-america/
6. Samborska, Katarzyna, Wasilewska, Aleksandra, Gondek, Ewa, Jakubczyk, Ewa and Kaminńska-Dwórznicka, Anna. "Diastase Activity Retention and Physical Properties of Honey/Arabic Gum Mixtures After Spray Drying and Storage" *International Journal of Food Engineering*, vol. 13, no. 6, 2017, pp. 20160320.
7. The Weights and Measures (Miscellaneous Foods) Order 1988 Reg. 3(3)
8. Brexit opportunities: regulatory reforms
   https://assets.publishing.service.gov.uk/government/uploads/system/uploads/attachment_data/file/1018386/Brexit_opportunities-_regulatory_reforms.pdf
9. The Weights and Measures (Miscellaneous Foods) Order 1988 Reg. 4(1)
10. Weights and Measures Act 1985 s36
11. Public Health England "Botulism: clinical and public health management" https://www.gov.uk/government/publications/botulism-clinical-and-public-health-management/botulism-clinical-and-public-health-management
12. https://www.qfood.eu/blog/1989-glass-in-baby-food/#:~:text=Rodney%20Whitchelo%2C%20a%20former%20Scotland,glass%20shards%20and%20razor%20blades.
13. https://www.food.gov.uk/business-guidance/regulated-products/novel-foods-guidance
14. https://www.tandfonline.com/doi/abs/10.1080/00218839.2017.1411178?journalCode=tjar20
15. Ann Chilcott in a talk on Colonsay Bees, WBKA, 11/11/2021
16. Heather Honey, Tony Jefferson, BBKA News 2013 212 (August) p19-21
17. CODEX STANDARD FOR HONEY https://www.fao.org/3/w0076e/w0076e30.htm
18. Collection and analysis of honey samples potentially contaminated with pyrrolizidine alkaloids from ragwort and borage, assessment of the stability of these compounds during storage of honey; FSA 2005 T01037
19. COMMITTEE ON TOXICITY OF CHEMICALS IN FOOD, CONSUMER PRODUCTS AND THE ENVIRONMENT COT Statement on Pyrrolizidine Alkaloids in Food https://cot.food.gov.uk/sites/default/files/cot/cotstatementpa200806.pdf
20. Sources of unpaletable or toxic nectar, Pam Hunter FLS, FRSocBiol, BBKA News, Special Issue, Advanced Husbandry 2018 p38
21. Organic Tracers from Asphalt in Propolis Produced by Urban Honey Bees, *Apis mellifera* Linn. Abdulaziz S. Alqarni & others

| | |
|---|---|
| | https://doi.org/10.1371/journal.pone.0128311 |
| 22 | https://echa.europa.eu/substance-information/-/substanceinfo/100.029.407 |
| 23 | https://www.legislation.gov.uk/uksi/2012/2619/contents |
| 24 | https://www.legislation.gov.uk/uksi/2005/1803/contents/made |
| 25 | https://www.bromley.gov.uk/leaflet/328933/6/402/ch |
| 26 | https://www.sciencedirect.com/science/article/pii/S1319562X17302073 |
| 27 | Regulation (EC) No 852/2004 of the european parliament and of the council of 29 April 2004 on the hygiene of foodstuffs, 2004 |
| 28 | Regulation (EC) No 852/2004 of the european parliament and of the council of 29 April 2004 on the hygiene of foodstuffs Chapter V https://www.legislation.gov.uk/eur/2004/852/annex/II/chapter/V# |
| 29 | https://www.youtube.com/watch?v=cCpr11OuYKI |
| 30 | https://www.bmj.com/content/bmj/2/5514/601.full.pdf |
| 31 | https://www.fsai.ie/news_centre/food_alerts/hilltop_honey_recall.html |
| 32 | https://www.imperial.ac.uk/news/215053/deaths-from-food-allergy-rare-decreasing/ |
| 33 | https://www.allergyuk.org/about-allergy/statistics-and-figures/ |
| 34 | The Honey (England) Regulations 2015 https://www.legislation.gov.uk/uksi/2015/1348/made |
| 35 | https://www.gov.uk/guidance/food-standards-labelling-durability-and-composition#honey |
| 36 | Honey A Comprehensive Survey: Eva Crane, 2020 |
| 37 | https://www.beeculture.com/processing-honey-a-closer-look/ |
| 38 | Effect of Storage and Processing Temperatures on Honey Quality: White, Jonathan W. ; Kushnir, Irene ; Subers, Mary H., Effect of Storage and Processing Temperatures on Honey Quality, 1964 3rd AOAC Europe, Eurachem Symposium, Brussels 3 March 2005: Biagio Fallico, Elena Arena, Mario Zappalà, Antonella Verzera |
| 39 | https://bbkanews.com/issues/view/152 |
| 40 | Utah Code § 4-5-502 https://le.utah.gov/xcode/Title4/Chapter5/4-5-S502.html |
| 41 | https://www.honeybeesuite.com/honey-pasteurization/ |
| 42 | https://www.rossendalefreepress.co.uk/news/haslingden-takeaway-boss-jailed-manslaughter-15384123 |
| 43 | https://www.bbc.co.uk/news/uk-england-36360111 |
| 44 | https://www.theguardian.com/uk-news/2021/dec/02/one-killed-31-left-ill-undercooked-shepherds-pie-pub-northamptonshire |
| 45 | https://www.bbc.co.uk/news/uk-scotland-tayside-central-38829175#:~:text=A%20Royal%20beekeeper%20who%20gave,be%20convicted%20of%20the%20charges. |
| 46 | https://www.nationalbeeunit.com/downloadDocument.cfm?id=1081 |
| 47 | https://www.honeyassociation.com/images/2012_ha_statement_on_gm_1_1.pdf |
| 48 | https://www.europarl.europa.eu/news/en/press-room/20140110IPR32407/parliament-clarifies-labelling-rules-for-honey-if-contaminated-by-gm-pollen |
| 49 | https://www.gov.uk/guidance/organic-food-uk-approved-control-bodies |
| 50 | https://www.theguardian.com/uk/2005/aug/21/foodanddrink.organics1 |

| | |
|---|---|
| 51 | https://www.theguardian.com/uk/2009/sep/22/director-jailed-fake-organic-food |
| 52 | The Weights and Measures (Specified Quantities) (Pre-packed Products) Regulations 2009 Reg. 5(2) |
| 53 | The Weights and Measures (Food) (Amendment) Regulations 2014 |
| 54 | Weights and Measures (Packaged Goods) Regulations 2006 |
| 55 | https://irna.fr/Honey-in-the-pyramids.html |
| 56 | https://www.ncbi.nlm.nih.gov/pmc/articles/PMC2271776/ |
| 57 | Webinar: The Global Honey Supply Chain 19 January 2022 The Government Chemist https://www.youtube.com/watch?v=IDjLF97Relo |
| 58 | https://www.researchgate.net/publication/230135131_Prediction_of_honey_shelf_life |

# Alphabetical Index

## A
| | |
|---|---|
| Acetic Acid | 13 |
| acidity | 26, 32, 66, 67, 77, 88, 94 |
| additives | 19, 61 |
| allergen | 23, 24, 49, 55, 56 |
| allergies | 37, 55 |
| anti-microbial | 32 |
| aspartame | 24 |
| Authority | 31 |
| Average System | 86 |

## B
| | |
|---|---|
| baker's honey | 60, 62, 63, 64, 66, 75, 94, 95, 96 |
| Bee Bread | 31 |
| beeswax | 11, 31, 33, 36, 37, 60, 71, 86 |
| Beeswax melts | 33, 37 |
| best before | 22, 83, 88, 89 |
| BHIPA | 26, 82 |
| Bisphenol A | 19, 47 |
| botulism | 26, 88 |
| Botulism | 26 |
| BPA | 19 |
| Brexit | 22, 59, 82 |
| British Honey Importers and Packers Association | 26, 82 |

## C
| | |
|---|---|
| cadmium | 19, 77 |
| caffeine | 24 |
| candles | 37, 38 |
| carcinogenic | 11, 19, 47, 54 |
| Caveat | 89 |
| colourings | 24 |
| comb | 18 |
| comb honey | 36, 60, 86 |
| compound food | 23, 32, 35, 43, 54, 55, 63, 71, 87 |
| cosmetic | 33, 38 |
| cosmetics | 33, 38 |
| cup and fork | 16, 18, 19, 47, 49 |
| cup and fork symbol | 16, 18 |
| cut comb | 18, 36, 46, 60, 86 |

## D
| | |
|---|---|
| diastase | 16, 66, 67, 77, 88, 89, 96 |
| Diterpenoids | 33 |
| Drones | 31 |
| due diligence | 16, 26, 44, 82, 83 |
| durability | 22, 35, 83, 87, 88 |

## E
| | |
|---|---|
| Egypt | 87 |
| electrical conductivity | 64, 66, 77, 95 |
| endocrine disruptors | 19 |
| enforcement | 69, 70, 78 |
| Extracting Rooms | 15 |
| extractor | 18, 32, 46, 47, 52, 57, 58 |

## F
| | |
|---|---|
| FCM | 46, 47 |
| feeding | 11, 12, 50, 75, 81, 84 |
| fermenting | 22, 66 |
| floppy baby syndrome | 26 |
| food business | 19, 27, 28, 41, 43, 44, 48, 49, 70, 71, 73, 78, 80 |
| Food business | 20, 28 |
| food business operator | 19, 20, 41, 70, 78, 80 |
| food contact material | 13, 16, 17, 18, 19, 33, 37, 45, 46, 47, 49, 57, 58 |
| food contact materials | 13, 17, 18, 19, 37, 45, 46, 47, 49, 57 |
| food handler | 15, 48, 49 |
| food rooms | 15, 39, 41, 44 |
| Food Standards Agency | 12, 54 |
| Food Standards Authority Agency | 31 |
| for food use | 33, 37, 47, 49 |
| free acid | 66, 96 |
| FSA | 14, 21, 24, 31, 32, 48, 49, 57, 58, 59, 64, 76 |

**G**

| | |
|---|---|
| geographical | 34, 64, 75 |
| glaze | 18, 19 |
| Glucose Oxidase | 32 |
| Government stamped | 87 |
| Government Stamped scales | 85 |
| gross weight | 22, 54, 87 |

**H**

| | |
|---|---|
| HACCP | 16, 27, 81 |
| harvest | 13, 14, 32, 36, 50, 64, 68, 70, 75, 87 |
| Hazard Analysis and Critical Control Point | 27 |
| Hive Cleaners | 12 |
| Hive protection | 13 |
| HMF | 16, 26, 32, 66, 67, 68, 74, 77, 83, 88, 96 |
| Honey | 31 |
| honeycomb | 10, 11, 31, 32, 33, 35, 36, 46, 47, 60 |
| honey houses | 15 |
| honey mustard | 24, 87 |
| hydrogen peroxide | 26, 32 |
| hydroxymethylfurfural | 66, 67, 77, 96 |

**I**

| | |
|---|---|
| ingredients | 24, 33, 37, 54, 56, 63, 69, 70, 71 |
| Invertase | 88 |

**L**

| | |
|---|---|
| label | 19, 20, 21, 22, 23, 24, 25, 26, 35, 38, 39, 47, 49, 54, 55, 56, 62, 63, 64, 69, 73, 75, 76, 82, 83, 85, 87, 88 |
| labelling | 19, 21, 23, 24, 25, 35, 38, 39, 49, 55, 56, 64, 69, 75, 76, 85 |
| lead | 18, 19, 26, 47, 58, 64, 70, 73, 75, 78, 84, 85 |
| liquorice | 24 |

**M**

| | |
|---|---|
| mad honey disease | 33 |
| medicinal | 36, 38, 39 |
| medicinal claim | 39 |
| melts | 33, 36, 37 |
| Minimum System | 86 |

**N**

| | |
|---|---|
| Naphthalene | 11 |
| nett weight | 22, 87 |
| Novel Foods | 31 |

**O**

| | |
|---|---|
| organic | 23, 32, 61, 66, 69, 83, 84, 85, 94 |
| overnment stamped | 87 |
| oxalic acid | 13 |

**P**

| | |
|---|---|
| Packing | 18 |
| pasteurise | 20, 67 |
| PDB | 11, 78 |
| pesticide | 12, 13, 47, 74 |
| pharmaceutical | 33, 35, 39 |
| pollen | 12, 17, 31, 32, 33, 34, 35, 39, 61, 64, 65, 66, 69, 75, 82, 94 |
| polyols | 24 |
| primary production | 24, 71 |
| prime producer | 53 |
| prime products | 71 |
| propolis | 14, 31, 35, 50, 71 |
| Pyramids | 87 |

**Q**

| | |
|---|---|
| quantitative ingredients declaration | 24 |
| quid | 24, 50, 60, 63, 75 |

## R

| | |
|---|---|
| raw | 20, 54, 56, 68, 69, 70, 71, 95, 96 |
| raw honey | 68, 69 |
| recycling jars | 57 |
| refractometer | 63, 88 |
| register | 20, 38, 70, 71, 73, 74, 83 |
| registration | 24, 70, 71, 73, 74, 84 |
| reserved description | 20, 83 |
| retail | 22, 24, 25, 54, 55, 60, 61, 71, 72, 73, 74, 76, 83 |
| ripening | 17, 47 |
| royal jelly | 71, 31, 36 |

## S

| | |
|---|---|
| second-hand | 57, 58 |
| sections | 18, 36, 46, 59, 60, 86 |
| security seals | 26 |
| smokers | 10 |
| Soil Association | 84 |
| sterilise | 51, 57 |
| sugars | 23, 24, 32, 63, 66, 67, 69, 74, 94 |
| sulphur dioxide | 13 |
| supers | 11, 12, 14, 15 |
| sweeteners | 24, 69 |

## T

| | |
|---|---|
| Three Packers Rules | 86 |
| tinplate | 47 |
| tne | 48, 85, 86 |
| Tolerable Negative Error | 86 |
| Tonics | 12 |
| Treating the bees | 12 |
| Turpentine | 37 |
| tutin | 33 |

## U

| | |
|---|---|
| UK Approved Control Body | 83 |
| Upcycling | 58 |
| use by | 58, 87, 88 |

## V

| | |
|---|---|
| Varroa | 13 |
| venom | 31, 36 |
| veterinary medicine | 13, 39, 79, 81 |
| Veterinary Medicines Directorate | 39 |

## W

| | |
|---|---|
| water insoluble content | 66, 67 |
| Weighing Systems | 86 |
| wholesale | 54, 81 |
| wraps | 33, 36, 37 |

## X

| | |
|---|---|
| x height | 24 |

www.ingramcontent.com/pod-product-compliance
Ingram Content Group UK Ltd.
Pitfield, Milton Keynes, MK11 3LW, UK
UKHW050019141125
465057UK00008B/135